INSTITUTION OF CIVIL ENGINEERS

Forensic engineering
The investigation of failures

Proceedings of the second international conference on forensic engineering organized by the Institution of Civil Engineers and held in London, UK, on 12–13 November 2001

Edited by B. S. Neale

ThomasTelford

Conference organized by Conference Office of the Institution of Civil Engineers on behalf of the Structural and Building Board.

Technical Committee:
>   Brian Neale, *Health & Safety Executive, UK* (Chairman)
>   Gerlando Butera, *Nabarro Nathanson*
>   Dr W Gene Corley, *Technical Council on Forensic Engineering, ASCE*
>   Michael Neale, *Neale Consulting Engineers Ltd*
>   Eamon O'Leary, *Consultant*
>   Professor Howard Wright, *University of Strathclyde*

International Scientific Committee (* = invited)
>   Mr Mohammad Ayub, *Occupational Safety & Health Administration, USA*
>   Mr Kimball J Beasley, *Wiss, Janney, Elstner Associates, Inc, USA*
>   Dr Kenneth Carper, *Washington State University, USA*
>   Professor Kees D'Angremond, *TU Delft, The Netherlands*
>   Dr Milos Drdácký, *Institute of Theoretical and Applied Mechanics, Academy of Sciences of the Czech Republic*
>   Professor Haig Gulvanessian, *Building Research Establishment, UK*
>   Dr Peter Ho, *Incode Pty Ltd, Australia*
>   Mr F S H Ng, *Paul Y. – ITC Construction Holdings Ltd, Hong Kong, ROC*
>   Dr Ilias Ortega, *Ortega & Kanoussi Technologies, Switzerland*
>   Mr Vivian Ramsey*, *Keating Chambers, UK*
>   Professor Rob Shepherd, *Earthquake Damage Analysis Corporation, USA*
>   Professor Sam Thorburn, *UK*
>   Professor Iris Tommelein*, *University of California at Berkeley, USA*
>   Professor Jonathan Wood, *Structural Studies & Design, UK*

Co-sponsored by
>   **The American Society of Civil Engineers**   **Institution of Mechanical Engineers**
>   **Institution of Structural Engineers**   **Health and Safety Executive, UK**

Published by Thomas Telford Publishing, 1 Heron Quay, London E14 4JD. www.thomastelford.com

First published 2001

A catalogue record for this book is available from the British Library

ISBN: 0 7277 3094 0

Printed and bound in Great Britain

Editor's preface

A warm welcome is extended to all delegates attending this Second International Conference on Forensic Engineering organized by the Institution of Civil Engineers. Welcome also to readers who were not able to attend. The international flavour of the conference extends from Hong Kong to Mexico and the USA, and from South Africa to Holland and the UK.

For this year's conference, we welcome the American Society of Civil Engineers (ASCE) as a co-sponsor, which joins in that role the UK-based, and internationally influential, Institution of Structural Engineers, Institution of Mechanical Engineers and the Health and Safety Executive.

The subject of this conference is the investigation of different types of failures of constructed facilities and learning from investigations of failures. It is aimed at and those who may be involved in some way, such as civil, structural, and mechanical engineers and architects. In particular, it is aimed at investigators and those who may have some influence in preventing failures in the future, both in the short term and also the medium and longer terms through design, construction, operation, assessment and maintenance of facilities. It will also be of interest to those who employ such professionals, including owners or occupiers of the constructed environment, facilities managers and those in the legal professions.

The conference is being held just two months after the attacks in the USA on the Pentagon in Washington DC, and the World Trade Center in New York which resulted in the subsequent progressive collapse of the twin towers. It is a time when the events are still very much in people's minds. It is a time also when the strengths and safety of buildings have come into focus, with much public discussion. Professional engineers in particular are making informed contributions to that debate. Structural safety is a focus of this conference, and many papers discuss levels of safety in a variety of contexts.

The papers in this volume cover a variety of subjects, such as: case studies of failures, ranging from catastrophic to serviceability failures, including mechanical engineering, structural engineering (both macro and micro) and geotechnical engineering; failures in a variety of structures (such as different types of buildings and bridges); failures in a variety of materials (such as steel, concrete, timber and masonry); claddings and fixings; failures due to a variety of initiators (including climatic and impact); investigation techniques and tools; presentation of evidence, for example for insurance, civil and criminal purposes; responsibility and culpability for failures; learning from failures; acceptable factors of safety, and the role of codes, including a new parallel development in both Europe and the USA.

A subject close to forensic practitioners, of whatever discipline, is the need to be seen as credible and professional, with their opinions and outputs taken as sound, appropriate and reliable. The introduction to the conference, through the Keynote Address (not included in these proceedings), considers the need for "convincing credibility" and includes descriptions of the background to and operation of the Council for the Registration of Forensic Practitioners (CRFP). This is a relatively new body, based in the UK, which is creating a register of practitioners, who will have to undergo scrutiny with respect to their credibility as experts before being allowed onto it. The aim is to help both forensic practitioners and those who employ them. Initially, the CRFP will be involved with forensic scientists.

This 2001 conference is building on the success of the first conference, held on 28–29 September 1998, by recognizing the widening interest in the topic. Thus the broader sponsorship has been complemented by a call for papers, rather than by inviting papers, hence each paper has been subject to a review process.

As Chairman of the conference organising committee, I should like to thank all the speakers and delegates, the co-sponsors, the Institution of Civil Engineers conference office (and in particular Barbara O'Donoghue and Penny Ryan) for their work with this conference, Thomas Telford Publishing and also Gerlando Butera, Dr Gene Corley, Eamon O'Leary, Michael Neale and Professor Howard Wright for their work on the Organizing Committee.

Brian S. Neale
Health and Safety Executive
Chairman of the Conference Organizing Committee

Contents

Effects of structural integrity on damage from the Oklahoma City, USA bombing

DR. W. GENE CORLEY, Senior Vice President
Construction Technology Laboratories, Inc., Skokie, Illinois, USA
ROBERT G. SMITH, Technical Consultant
Erico, Inc., Solon, Ohio, USA
LOUIS J. COLARUSSO, Senior Development Engineer
Erico, Inc., Solon, Ohio, USA

ABSTRACT
Starting in 1989 (Ref. 1), the ACI Building Code began requiring integrity reinforcement at the perimeters of all buildings and at all slab column intersections. The Commentary to Section 7.13 states that the purpose of this reinforcement is

> "to improve the redundancy and ductility in structures so
> that .....damage to major..... element.....be confined to a
> relatively small area and the structure will....maintain
> overall stability."

This paper reviews requirements for structural integrity reinforcement after 1989, describes a case study of blast-damage to the Murrah Federal Building in Oklahoma City, and shows how the use of full capacity butt splices could have greatly reduced the casualties in that blast.

INTRODUCTION

Required Integrity Reinforcement
Since 1989, requirements for structural integrity reinforcement have been listed in Section 7.13 of the ACI Building Code (Ref. 1). In general, Section 7.13 requires one-sixth to one-quarter of the reinforcement in perimeter beams to be continuous about the building. In addition, Section 13.4.8.5 requires two bottom bars in each direction be carried through columns continuously or anchored within the columns.

The purpose of integrity reinforcement for both spandrel beams and slab column intersections, is to provide a small capacity, even after failure of a column or slab at any one location.

In Chapter 21, toughness is required throughout the structure for seismic resistance. to obtain required toughness, Section R21.3.2 states the following:

> "Lap splices.....are prohibited at regions where flexural
> yielding is anticipated because.....splices are not reliable
> under.....cyclic loading into the inelastic range."

Oklahoma City Bombing
In 1995, the Murrah Federal Office Building in Oklahoma City was heavily damaged by a terrorist bomb blast (Ref. 2). A forensic investigation of damage to the Murrah Building disclosed that failure occurred in three columns supporting a transfer girder on the north

side of the building. One of these columns was destroyed by brisance or shattering while the adjacent two failed in shear.   After losing support from the three columns, approximately 50 percent of the floor area of the building collapsed, producing a large number of casualties. Details of the forensic investigation are reported in Ref 2.

A photo of the Murrah Building prior to the terrorist attack is shown in Fig. 1. The transfer girder supported on four intermediate columns can be seen just below the glass curtain wall.
Fig. 2 shows a drawing of the building north elevation with the explosive-laden truck parked in the street beside it. Also shown in the figure is the resulting crater. Fig. 3 shows a plan view of the crater and location of the column nearest the truck. As reported in Ref. 2, a bomb having an explosive power equal to approximately 4,000 lbs. of trinitrotoluene (TNT) was detonated in the truck.

Analysis of damage in the field and calculation of effects of the explosion determined that the nearest column, G20 as identified in Fig. 4, would have been destroyed by brisance or shattering. Approximately 40 ft each side of the column that was destroyed, adjacent columns were found to have exceed their shear strength, but been approximately at their flexural capacity. As a result, the adjacent columns failed in shear, thereby leaving the spandrel beam without support for distance of approximately 160 ft.

As indicated in Fig. 5, three of the transfer girder top bars were continuous but none of the bottom bars were. A similar pattern was present in spandrel girders above the third floor as shown in Fig. 5. Consequently, there was no effective integrity reinforcement in this building.   It must be noted that at the time the building was built, no integrity reinforcement was required. Analyses show that this building met or exceeded code requirements in every way that was checked.

Ref. 2 indicated that shear failures of two of the three columns could have been prevented if only a small amount of hoop steel had been provided. It is also possible that the column nearest the bomb could have been saved if full confinement reinforcement had been provided. However, for purposes of this discussion, it will be assumed that the column nearest the truck would be destroyed by brisance.

In Ref. 2, an analysis is made with one column removed. The three mechanisms that could develop after removal of one column are shown in Fig. 6. Calculations indicate that Mechanism 2 results in a capacity of approximately 60 to 70 lbs. per sq ft floor above the spandrel beam. This is significantly less that the approximately 110 lbs. per sq ft of self-weight plus live load that existed in the building. Consequently, removal of any one column supporting the as-built transfer girder would cause failure of all of the floors above over a length of about 80 ft.

FULL CAPACITY MECHANICAL BUTT SPLICES

Reduction in Loss
In the 1999 ACI Code, (Ref.3) full capacity mechanical butt splices were recognized for the first time. Classified as Type 2 mechanical splices in Section 21.2.6.1(b), they are required to "develop the specified tensile strength of the spliced bar." ACI 318-99 permits use of Type 2 splices at any location, including hinging regions. Consequently, Type 2 full capacity mechanical butt splices can be used to connect integrity reinforcement and improve the blast and/or seismic resistance of concrete structures.
If integrity reinforcement in the spandrel beams of the Murrah Building was maximized by making all of the bars continuous with full capacity mechanical butt splices, significant increase in the capacity of Mechanism 2 of Fig. 6 would be realized. Using Mechanism 2

with 100 percent of the bars spliced with full capacity mechanical butt splices, calculated capacity of the building would be at least 140 lbs. per sq ft. Unit weight of the building, including some live load, is approximately 110 lbs. per sq ft. Consequently, even if one column had been removed the building would not have collapsed.

Although collapse of the building frame would have a high probability of being prevented by making all of the reinforcement in spandrels continuous, several of the floors would be destroyed as a result of brisance. Consequently, the reduction in catastrophic damage to the building would not be 100 percent, but is estimated to be approximately 80 percent.
As indicated in Ref. 2, almost all of the casualties were the result of building collapse, not air blast. Consequently, if building collapse is reduced by 80 percent, casualties would be reduced by a similar amount.

Cost Analysis
To determine what the additional cost might be if full capacity mechanical butt splices had been used in the Murrah Building, a review of cost studies done by Cagley and Associates and reported in Ref. 4 was made. Fig. 7, courtesy of ERICO, shows their full capacity butt splices being installed on bars approximately the size of those used in the Murrah Building. Fig. 8 shows a close-up of an ERICO full capacity butt splice. As can be seen, these splices are relatively small in size and not very difficult to install.

Cost analyses done by Cagley and Associates (Ref. 4) show that the increased cost of mechanical butt splices over lap splices in a 12-story parking deck was less than ¼ of 1 percent of the cost of the building. In an office structure, such as the Murrah Federal Building, the cost of mechanical splices would be even a smaller percentage of the total cost. Considering that use of full capacity butt splices could be limited to only that portion of the building exposed to the street where the bomb was detonated, estimated total cost differences are approximately 1/8 of the total cost of the building. This cost difference is insignificant.

SUMMARY
This paper discusses the potential for using mechanical full capacity butt splices to increase blast and earthquake resistance of buildings. The need for integrity reinforcement to reduce damage in a building when unanticipated intense loads destroy a single element is discussed. A prohibition against lap splices in hinging regions of buildings designed to resist earthquakes is noted.

Terrorist bombing of the Murrah Federal Building in Oklahoma City is used as a case study. It is shown that without integrity reinforcement, strengthening of columns could only reduce damage by about 50 percent. However, if full capacity mechanical butt splices had been used to make all of the spandrel beam reinforcement continuous, collapse of the building would have been reduced by an estimated 80 percent with a similar reduction in casualties. The estimated additional cost to provide this continuous reinforcement is approximately 1/8 of 1 percent, much less than the variation in estimating the cost to construct the building.

Fig. 1 Murrah Building Prior to Blast.

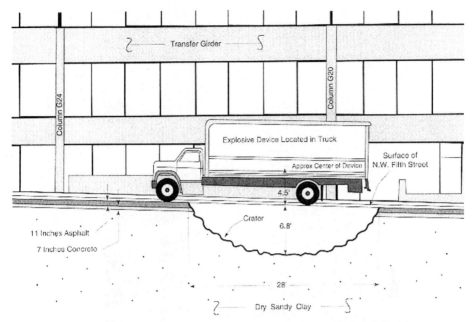

Fig. 2   Approximate Dimensions of Crater at North Face of Murrah Building.

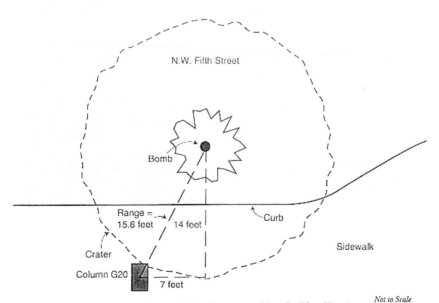

Fig. 3   Proximity of Column G20 to Location of Bomb (Plan View).

Fig. 4   Damage to North and East Sides of Murrah Building.

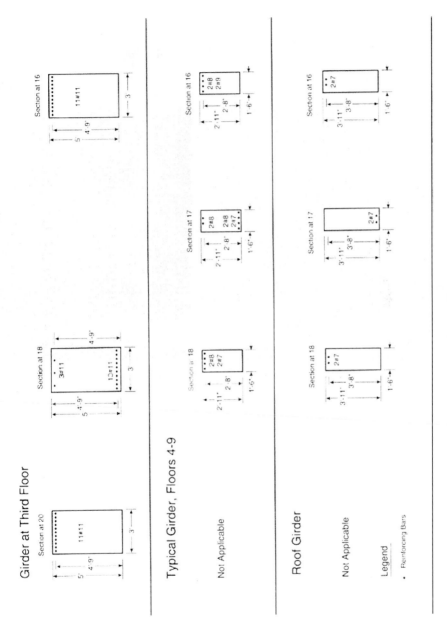

Fig. 5 Cross-Sectional Distribution of Girder Reinforcing Bars.

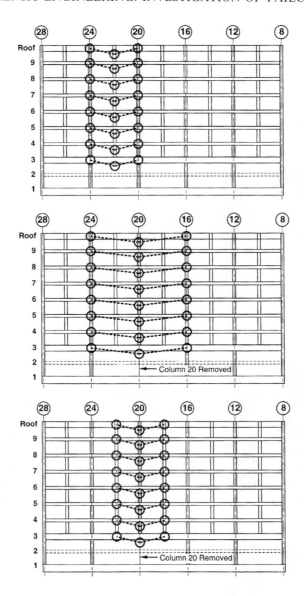

Fig. 6   Possible Mechanisms with No Columns Removed and One Column Removed.

Fig. 7   Installation of Mechanical Splices.

Fig. 8   Full Capacity Mechanical Butt Splice Installed in Reinforcing Cage.

REFERENCES

1. American Concrete Institute, *Building Code Requirements for Reinforced Concrete* (ACI 318-89), ACI Committee 318, Detroit, Michigan, 1989.

2. Corley, W. G., Sozen, M.A., Thornton, C.H., and Mlakar, P.F. (1996). "The Oklahoma City Bombing: Improving Building Performance Through Multi-Hazard Mitigation." *FEMA Bulletin 277*, Federal Emergency Management Agency, Washington, D.C.

3. American Concrete Institute, *Building Code Requirements for Structural Concrete* (ACI 318-99), ACI Committee 318, Farmington Hills, MI, 1999.

4. Cagley, James R., and Apple, Richard, "Economic Analysis—Mechanical Butt Splices vs. Lap Splicing, ERICO, Inc., Solon, Ohio, 5 pp.

CONVERSION FACTORS—INCH-POUND TO SI (METRIC)*

To convert from	to	multiply by
	Length	
inch	millimeter (mm)	25.4E†
foot	meter (m)	0.3048E
	Area	
square inch	square centimeter (cm^2)	6.451
square foot	square meter (m^2)	0.0929
	Volume (capacity)	
cubic inch	cubic centimeter (cm^3)	16.4
cubic foot	cubic meter (m^3)	0.02832
	Force	
kilogram-force	newton (N)	9.807
kip-force	newton (N)	4448
pound-force	newton (N)	4.448
	Pressure or stress	
	(force per area)	
kilogram-force/square meter	pascal (Pa)	9.807
kip-force/square inch (ksi)	megapascal (MPa)	6.895
newton/square meter (N/m^2)	pascal (Pa)	1.000E
pound-force/square foot	pascal (Pa)	47.88
pound-force/square inch (psi)	kilopascal (kPa)	6.895
	Bending moment or torque	
inch-pound-force	newton-meter (Nm)	0.1130
foot-pound-force	newton-meter (Nm)	1.356
meter-kilogram-force	newton-meter (Nm)	9.807
	Mass per volume	
pound-mass/cubic foot	kilogram/cubic meter (kg/m^3)	16.02
pound-mass/cubic yard	kilogram/cubic meter (kg/m^3)	0.5933
pound-mass/gallon	kilogram/cubic meter (kg/m^3)	119.8

* This selected list gives practical conversion factors of units found in concrete technology. The reference source for information on SI units and more exact conversion factors is "Standard for Metric Practice" ASTM E380. Symbols of metric units are given in parentheses.

† E indicates that the factor given is exact.

Ladbroke Grove Crashworthiness Accident Investigation

MR A.P.BRIGHT MEng, CEng, MIMechE, WS Atkins Consultants Ltd, Bristol, UK
DR N.KIRK BSc, PhD, CEng, MIMechE, WS Atkins Consultants Ltd, Bristol, UK
DR S.MANTEGHI PhD, CEng, MIMechE, SenMWeldI, WS Atkins Consultants Ltd, Bristol, UK
MR P.J.MURRELL MEng, CEng, MIMechE, WS Atkins Consultants Ltd, Bristol, UK

INTRODUCTION

On the 5th October 1999 a high speed Intercity train (HST) collided with a local train service (Class 165) at a closing speed of over 130 mph near to Ladbroke Grove junction, Paddington. Seven fatalities were sustained in the HST which was fabricated using welded mild steel sections, and 24 fatalities in the Class 165 train which was fabricated from welded aluminium extrusions.

This paper describes the key activities that were undertaken by WS Atkins Consultants Ltd on behalf of the Health and Safety Executive to investigate the structural and crashworthiness aspects of the accident. The aims of the crashworthiness investigation were to:

- Determine how the vehicles and associated structures had performed during the initial impact, and whether this had contributed to the number and distribution of fatalities.
- Verify that the vehicles were manufactured to their design specifications.
- Assess whether there were any gross differences in the crashworthy performance of steel and aluminium rail vehicles.

Extensive surveys of the accident scene and damaged vehicles were undertaken. In addition a full scale reconstruction of the structural members from the leading vehicles from each train was undertaken to determine how these structures had failed under the extreme impact conditions. From this data a detailed understanding of the accident dynamics and vehicle performance was obtained.

A comprehensive programme of material and component tests was performed to ensure the vehicles had been built to their design specification. These tests were also used to understand at a basic level the impact performance of the joining processes and materials used in each design, and how these properties contributed to the failure modes observed in the accident.

Due to the high closing speed and different configuration of the trains involved in the accident it was not possible to achieve a valid comparison of the impact performance of steel and aluminium rail vehicles using purely the evidence from the accident. In order to gain a fairer comparison an assessment programme was undertaken using finite element modelling techniques.

Forensic engineering: the investigation of failures. Thomas Telford, London, 2001.

CRASHWORTHINESS

The term 'crashworthiness' refers to the capability of a vehicle or component to provide occupant protection during potentially survivable collisions. Crashworthiness is guided by four distinct concepts:

1. Limiting to tolerable levels the impact forces that are applied to occupants.
2. Providing means for managing the energy of collision while at the same time maintaining adequate survival space for the occupants.
3. Protecting the occupants within the vehicle survivable space during the collision. This may be achieved by minimising the likelihood of passenger ejection and by minimising the likelihood of intrusion from debris.
4. Protecting the occupants from post-crash hazards, primarily fires.

All the vehicles involved in the accident were manufactured prior to the introduction of modern crashworthiness design standards. They were designed only to meet static proof and fatigue load criteria. Therefore, none of the vehicles accommodated modern crashworthy features such as crumple zones, shear out couplers or anti-climbers.

PUTTING THE ACCIDENT INTO CONTEXT

Due to the high speed of the trains in the accident, the energies involved were very high. The HST had a kinetic energy of $_c$400MJ and the Class 165 had $_c$30MJ of kinetic energy.

Modern vehicles are designed for the most likely type of collision. Historically, these have been end on impacts at closing speeds below 40mph. Crush zones in modern carriages are therefore designed to absorb the energies associated with these types of impacts. Typical absorption levels range between 2 to 4 MJ per vehicle.

The energies involved in the accident were considerably in excess of the current design criteria. It would therefore be expected that a modern crashworthy vehicle would have sustained significant structural damage in a similar impact scenario. In such high-speed accidents it is the ability of the vehicle to maintain the integrity of the passengers survival space, rather than to absorb all the energy, that becomes the most important factor for occupant survivability.

OVERVIEW OF THE DAMAGED VEHICLES

Class 165

The Class 165 train consisted of a three-car diesel multiple unit. The Class 165 bodyshell is a stiff monocoque structure primarily fabricated from 6005A extruded aluminium alloy components with a T6 heat treatment. The extrusions are joined using 4000 series weld material.

The main structural members of the bodyshell consist of a series of longitudinal aluminium alloy extrusions at floor and roof level, with vertical pillars at the ends and also between the positions of doors and windows. The main longitudinal extrusions are the solebars at floor level, the floor planks and the cantrails at roof level. At vehicle ends there are major vertical pillars either side of the gangway and at the outer corners of the vehicle. The sides of the vehicle, between the solebar and cantrail, are also fabricated from extrusions. The vehicles are supported on steel bogies and have the aluminium fuel tanks suspended beneath them.

The **leading vehicle** fragmented into many pieces with many fractures in the aluminium sections and plating. Debris from the leading vehicle was scattered from the point of impact towards the Paddington end of the crash site, over a distance of approximately 190m. To aid the investigation process a 2D reconstruction (Figure 1) of the leading vehicle was subsequently undertaken in Crewe.

Figure 1: 2D Reconstruction of the Leading Class 165 Carriage

In the **centre vehicle**, fractures were observed in the following locations:

- Butt welds at the base of gangway and corner pillars.
- Welds between the gangway pillar and upper cross beam.
- Welds between the sidewall panels (upper solebar) and the lower solebar at floor level.
- Welds between the side or roof panels, slip joint and lower cantrail at roof level.
- Welds at the deadlights associated with window openings.
- Welds between the corner pillar and front headstock.

The **rear and final carriage** had fallen over onto its side and was mostly undamaged. There were two significant penetrations in each bodyside where they had been punctured by massive items during the accident.

Twenty-four fatalities were sustained in the Class 165 train. Twenty-one were from the leading vehicle of which only one was not impact related. There was one impact related fatality from the trailing vehicle, with the other two from the leading end of the centre vehicle.

High Speed Train (HST)

The HST consisted of eight Mark 3 coaches with Class 43 power cars at either end.

Class 43 Power Car

The Class 43 power car is fabricated from mild steel components, welded together to form a robust leading end underframe and monocoque trailing end structure. The main structural features of the Class 43 are summarised below:

- The leading end of the Class 43 is extremely robust at underframe level.
- The leading part of the superstructure of the Class 43 cab is fabricated from GRP and wood. It has no significant structural strength above the underframe. The first structural element above floor level is the door pillar behind the driver's compartment.
- The trailing end is of similar construction to that of a Mark 3, consisting of a framework fabricated from welded steel box sections.

The leading power car of the HST came to rest in the fence on the South side of the track approximately 92.5 m from the estimated point of impact. The vehicle was upright and had suffered significant damage at both the leading and trailing ends. The front end of the HST power car had disintegrated (Figure 2). The bolster to solebar weld on one side, and the weld between the solebar and the front headstock on the other side had broken completely. The central portion of the power car was largely intact, although the trailing end had been partially overridden by Coach H.

Figure 2: Leading HST Power Car

Mark 3 Coaches

The Mark 3 coach is fabricated from mild steel components, welded together to form a monocoque structure. The main structural features are summarised below:

- The underframe consists of two pressed steel solebars running down each side of the vehicle. Smaller pressed steel sections and corrugated plates form the floor and other longitudinals and transverse members within the floor. The underframe is strengthened in the regions of the bolster and dragbox to accommodate the higher loads in these areas.

- The sides and roof are fabricated from welded 2mm thick pressed steel sections, with a cantrail running along the interface between the roof and side sections.
- The intermediate ends consist of a framework fabricated from welded steel box sections.

Only the first two coaches sustained significant structural damage. Coach H had jack-knifed by 180 degrees. Both ends of the vehicle were significantly damaged by the initial impact and the carriage had been burnt out by fire following the collision.

Coach G had fallen onto its side, coming to rest approximately 129m from the impact point. The bodyshell was largely intact with the exception of some localised impact damage to the leading end.

In addition to the HST driver, six fatalities were sustained in Coach H, five of which were impact related. All the impact related fatalities were from the first 2m of the leading end of Coach H. Several of the occupants had been standing in the vestibule and toilet area prior to the impact.

VEHICLE INSPECTION
Approximately eighty samples were identified and from the Class 165 and HST vehicles for further investigation. The majority of these were taken from the Class 165 vehicles. Samples were taken from all of the main types of fracture and also from undamaged regions of similar locations for comparative purposes.

Class 165
In general, the observed quality of welding on the Class 165 vehicles in terms of internal soundness and freedom from weld cracking was satisfactory and appropriate for the type of fabrication.

Metallurgical studies indicated some porosity, most of which was within an acceptable range for the class of welding. There were a number of examples that exhibited extensive 'pipes' of porosity originating from the weld root. This is typical of a situation, not unique to aluminium, where trapped air in a partial penetrated weld breaks through to the surface.

A large number of welds showed only partial penetration. In the majority of the cases was done intentionally to ease design, manufacturing and assembly practicalities, rather than an issue of quality of construction.

Structural failure was predominantly associated with, or initiated at, weldments. Such failure was either through the weld metal or from the edge of the weld. There was no evidence of pre-existing fatigue cracks or any influence of fatigue in the failures. They originated at discontinuities in section at either the root or the toe of the weld. Any failures in parent material outside the immediate weld area could be explained by the fact that the overall design of the connection made the failure location the region of lowest cross-sectional area.

A number of welds were small in cross section and were only partially penetrated. In addition, some of the attachment welds around the corner pillars and door pillars barely penetrated. These were difficult manual welds, probably with poor access. So, high quality weld deposits in these locations would be difficult to achieve consistently.

The metallurgical characteristics of the Al-Mg-Si heat treatable alloys (their high ratio of proof strength/ultimate tensile strength, and their sensitivity to thermal effects of welding) make the issue of design for crashworthiness in these materials particularly critical.

The tensile and hardness tests from the metallurgical investigations were consistent with the welding procedure documents. They indicated that the weld metal under-matched the strength of the parent material away from the welds. They also confirmed the presence of a soft band of parent material alongside the welds where 6000 alloys had been used. Fracture toughness tests indicated low crack tip opening displacement values in the parent 6000 material. However, no extensive fractures occurred in the parent material because they occurred preferentially in the lower strength heat affected zone (HAZ) and weld metal regions.

In a number of cases, welds had been designed to have a throat less than the thickness of adjacent parent material. It is inevitable that in severe overload conditions, failure will occur at the weakest position in the load path. With the combination of reduced strength of weld metal and of parent material close to the weld, taken with the reduced cross sectional area in some cases, failure at the welded joints is inevitable under such loading.

The fractures at or near the welded connections vehicles were ductile overload failures. However, because of the under-matching strength of the joints, plasticity was confined to the local regions of the joints and only minimum amounts of energy were absorbed. It seems essential that, for optimum behaviour under impact conditions, joints and connections which are part of the energy absorption path should not under-match the parent material in strength.

High Speed Train (HST)
The welded connections in the Class 43 power car and Mark 3 coaches behaved much better than those in the Class 165 vehicles. It is noteworthy that in this case, the welds did not under-match the parent material in strength. There was plenty of evidence of global deformation of the parent material with relatively few weld failures.

PERFORMANCE ASSESSMENT
Due to the high closing speed and different configuration of the trains involved in the accident it was not possible to achieve a valid comparison of the impact performance of steel and aluminium rail vehicles using purely the evidence from the accident. In order to gain a fairer comparison an assessment programme was undertaken using finite element modelling techniques.

Finite element models of comparable steel and aluminium vehicles were developed and a range of impact scenarios was assessed using LS-DYNA (Figure 3). The models were validated, where possible, against vehicle and component test data.

The Class 165 was chosen for the aluminium design and the Class 321 as the "equivalent" steel design. These vehicles are used for similar duties and are of similar size and generation, both being manufactured in the mid to late 1980s. The Class 165 vehicle is a very stiff monocoque design fabricated from 6000 series aluminum extrusions joined by under-matched welds. The Class 321 is fabricated from mild steel sections joined by over-matched welds. Neither the Class 165 nor the Class 321 vehicles were designed to modern crashworthy design standards.

For the purposes of the assessments each model was impacted with a representation of the front end of a HST power car at a variety of speeds and offsets.

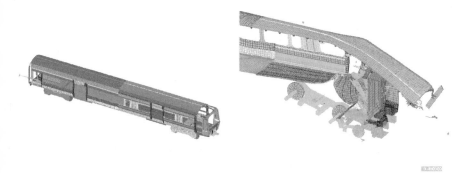

Figure 3: Examples of the Finite Element Impact Modeling

Historically, the primary cause of injury and loss of life in a rail crash has been the loss of survival space or ejection from the vehicle. Fatalities resulting from secondary impacts within the vehicle or due to debris intrusion are generally less frequent.

Based on the results of the FE analyses, the aluminium vehicle was found to offer better projection of passenger survival space. This was due to the high strength and stiffness of the double skinned extruded bodyshell. The main risk was that the mode of failure in the model was dominated by fracture of welds. This could lead to a catastrophic loss of structural resistance to collapse once failure is initiated. It was also noted that the higher strength of the aluminium vehicle was likely to lead to preferential damage to other vehicles. This may result in protecting the occupant of the aluminium vehicle at the expense of other railway users.

High decelerations, missile generation and debris intrusion, were also found to be characteristics of the collapse of the aluminium vehicle. These features are likely to cause more secondary injuries.

The steel design performed less well due to the relative weakness of the structure and its tendency to deform preferentially in the passenger survival space. The fact that this particular design of vehicle did not collapse sequentially from the end was concerning. In a good crashworthy design plastic deformation should begin preferentially at the vehicle ends, which have minimum occupant densities, and then subsequently progress towards the centre of the vehicle.

However, the mode of failure for the steel design was dominated by plastic deformation, rather than fracture, which reduced the risk of sudden structural failure, missile generation and occupant ejection.

In summary, if this aluminum design could be made less aggressive to other vehicles and less prone to fracture then it would exhibit good crashworthiness characteristics. For this steel vehicle, the main concern was the prevention of preferential deformation of the passenger compartment.

It should be noted that only rakes of two vehicles colliding with the Class 43 were considered in the analysis. The analyses could not be fully generic due to the wide range of different designs or both aluminium and steel rolling stock.

CONCLUSIONS

The conclusion from our investigations was that both steel and aluminium vehicles have their own strengths and weaknesses in terms of crashworthiness. Importantly there is no significant reason why steel and aluminium structures could not be designed to be equally crashworthy.

Mild steel is more ductile than 6000 series aluminum. Mild steel rail vehicles can undergo large deformations before any material fracture occurs. For dynamic situations, the higher ductility of the steel structure resulted in lower decelerations and less propensity for the bodyshell to fragment. This reduced the likelihood of opening up of the passenger compartment, occupant ejection, and the generation of jagged edges and missiles.

The higher stiffness of the aluminum structure resulted in higher decelerations, but with a more robust collapse mode. As result of the higher loads, the aluminum vehicle typically absorbed more energy. The under-matching of welds resulted in bodyshell fragmentation at impact speeds above 15mph. This resulted in opening up of the passenger compartment along longitudinal weld lines. This increased the risk of sudden loss of structural resistance to collapse, occupant ejection, debris intrusion and the generation of large jagged edges and missiles.

If the Class 165 could be made less aggressive to other vehicles and less prone to fracture then it would exhibit good crashworthiness characteristics. For the steel design considered, the main concern was the prevention of preferential deformation of the passenger compartment. This is considered to be a poor crashworthy design characteristic, though this is not thought to be a generic feature of all steel vehicles.

ACKNOWLEDGMENTS

The authors would like to thank the following organisations and individuals for their contribution to this work: Her Majesty's Railway Inspectorate, TWI, Health and Safety Laboratory, Prof F.M Burdekin, Prof J.H. Rogerson and the Bombardier Crewe works.

A staged forensic approach to resolve geotechnical engineering claims

EUR ING ROGER P THOMPSON BSc, MSc, CEng, FICE, FGS, MAE
EDGE Consultants UK Limited, Manchester, UK

INTRODUCTION

Failures involving ground engineering may sometimes be dramatic, for example when landslides occur, or may be rather more insidious, such as progressive settlement of developments built on backfilled quarries. Rigorous examination of the available records and in depth physical investigation is often required to establish the likely cause of the failure. Early mutual acceptance of the technical aspects by the parties involved can lead to an agreement that avoids costly litigation. The paper reviews six different case histories of geotechnical engineering failures. Each matter was resolved at progressively later stages with the final example following a High Court judgement. It demonstrates the cost saving potential that is available and can be realised by early resolution. Comparison is made with the erosion of cost saving potential and increased costs that result from allowing a dispute to continue until a trial takes place. The advantages of detailed forensic examination combined with a staged approach are shown to bring added value to the resolution of claims.

CASE HISTORIES

Landslide in Northwest England (1)

In order to form an access road to a site for a new development at a brewery, a cutting was made across a sloping valley side. Rainfall was reasonably heavy and the toe of the new cutting, after it had been made, was recognised by the contractor as being unstable. However, a walkover by the consulting engineer revealed that fissures had opened up about 100m up-slope from the cutting. It was evident that a relatively substantial landslide was in the process of occurring. Stabilisation works required a knowledge of the depth to the plane of sliding and the relevant groundwater pressures. This information could only be obtained by detailed borehole examination and groundwater monitoring. Temporary fragile stability was achieved by placing a toe berm. After investigations and monitoring lasting almost one year were complete a full understanding of the conditions controlling the slope movement was obtained. Remedial works involving a deep drainage trench were constructed to lower the groundwater pressures permanently and thereby prevent future slope movement. Some technical details of this project are presented in Leach and Thompson (ref 1) and a section through the landslide is given on figure 1.

Forensic engineering: the investigation of failures. Thomas Telford, London, 2001.

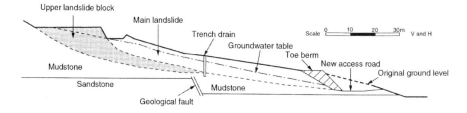

Figure 1. Cross section through landslide in Northwest England

From the outset, the developer client appreciated that all efforts should be focused on understanding and resolving the problem. There was close co-operation between the contractor and the consulting engineer. It could have been maintained that the problem might have been foreseen. However the post-failure investigations identified particularly complex ground and groundwater conditions such that until slope movement occurred it would have been very difficult to predict the critical values of the key parameters. Although expenditure was incurred in stabilising the failed slope there would have been greater expenditure in constructing conservative stabilisation measures to the slope in its pre-failure condition. Accordingly there was little if any merit to be gained by the developer in making a claim for the recovery of the cost of the works. This exemplifies the situation where the maximum potential for cost saving on litigation was utilised with all the enthusiasm of the parties being targeted towards achieving an early technical and contractual resolution.

Subsidence in Northeast England (2)

During the construction of a golf-driving centre in the Northeast of England, there was a torrential rainstorm. Subsidence occurred which affected the partly constructed facility. Repairs were put in hand but before the works were completed there was a further rainstorm and a repeat of the subsidence. The developer decided to obtain additional advice on the nature of the site. Careful investigation established that the material which had previously been considered to be natural ground was in fact fill to an old limestone quarry. There appeared to be some conflict between the extent of the quarry as shown on historical records compared with that depicted on the geological map of the area. This is depicted on figure 2 which shows a cross-section through the site. Although washout arising from the rainstorm was a possible cause of the subsidence, it was deemed that the fill had undergone collapse compression. Piling remedial works were therefore introduced.

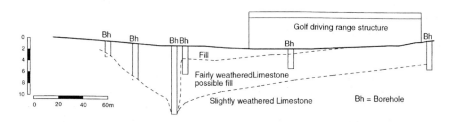

Figure 2. Cross section through site of backfilled Limestone quarry in Northeast England

Five possible defendants were involved and the developer decided to take action against three of them. Expert technical review led to the acceptance that the mostly likely cause of the subsidence was collapse compression. However there was a lack of agreement as to whether it could have been foreseen and whether there were shortcomings in the original investigation. Furthermore there appeared to be cost constraints imposed by the developer which might have restricted the scope of the original investigation and construction works.

Full litigation was considered but it was decided to try to achieve a resolution by mediation, before the stages such as disclosure of documents were reached. An advantage was that there was broad agreement on the technical cause for the subsidence. It was therefore a matter of debating the foreseeability and degree of responsibility that might be attached to the relevant parties together with the quantum. The mediation was carried out within one day. The differences between the parties were almost resolved on the day with the full resolution being achieved the next day. It ensured that the costly stage of document disclosure and examination was not reached. The example demonstrates that while some costs were expended on both legal and technical expert input, the potential for major cost savings compared to full litigation was realised by invoking mediation.

Subsidence in West Midlands of England (3)
Many years after houses were built on both sides of a road in the West Midlands, a number of the properties began to suffer settlement giving rise to structural cracking. Research revealed that the houses were located over backfilled sandstone quarries. Investigation demonstrated that the fill was predominantly loose sand with the groundwater level being close to the base of the old quarry. Repairs involving pile underpinning were put in hand before the degree of damage became excessive. Expert advice was sought regarding the cause of the settlement and why it should have arisen many years after the houses were built.

Some months before the damage had arisen, heavy lorries had trafficked the road in connection with a new housing development. Loose sand is susceptible to settlement on severe vibration. However, although the residents reported that their houses experienced vibration, the timing did not coincide with the damage. Excessive water was discharged from the site of the new housing development and flowed down the road past the existing houses. Again, detailed review showed that the two events did not coincide. Heavy rainfall occurred in the months immediately preceding the subsidence and initially appeared to be a probable cause. However examination of the rainfall records for 30 years revealed there had been other periods of high rainfall but that no subsidence had been reported. Shortly after the damage was observed, a major water main in the road was found to be leaking and was repaired. Records reported that the road was exhibiting damage for a number of weeks at the location of the leak before the leak was mended. Water from the leak could readily have inundated the loose sand fill causing collapse compression. The timing matched the occurrence of damage to the houses. Figure 3 shows the juxtaposition of the houses and the water main.

The insurers for the householders made a claim to recover the monies expended on the remedial works. The legal representatives recognised that disclosure of documents was necessary for the technical experts to review the full background to the matter. A valuable meeting of the experts was held at which the issues were discussed and some differences were

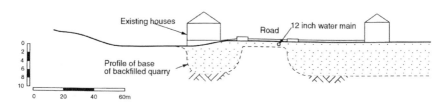

Figure 3. Cross section through site of backfilled Sandstone quarry in West Midlands

resolved. Expert opinion reports were prepared for exchange prior to the final preparation for a hearing. At this point a financial resolution between the parties was reached. Although quite considerable monies were expended by both the legal and technical expert parties, the potential of saving the cost of a court hearing and the inherent preparation was realised.

Landslide in West Midlands of England (4)

To provide a level area of sufficient size to accommodate a row of houses for a new development, it was decided to place fill on the upper part of a slope. The original natural ground level was masked by up to about 10m of ash fill as can be seen in figure 4. The newly created slope profile was initially stable and construction of the houses commenced. When they were partly completed there was a period of heavy rain. The brickwork to the houses began to show signs of cracking which progressively became more severe. The developer decided to seek independent technical advice.

Examination revealed a linear crack in the ground running parallel to and a few metres back from the crest of the slope. It was apparent that the complete slope had just begun to fail. A review of the historical records gave an indication of the possible thickness of fill while the published geological information established the likely natural stratigraphy. Boreholes were put down to allow sampling and testing and groundwater monitoring was undertaken. It was important to ascertain whether the materials had just suffered movement for the first time or whether they were undergoing a renewal of past sliding. Enquiries identified a nearby site where construction was in progress and where the same stratigraphy was present but not buried by fill. Examination of the ground in the sides of drainage trenches revealed a well-defined existing almost horizontal plane of discontinuity with a polished surface.

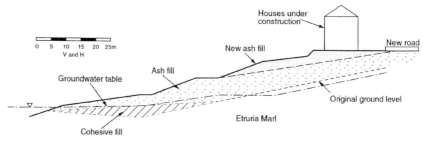

Figure 4. Cross section through landslide in West Midlands

It was concluded that this material had undergone past movement. Accordingly the slope failure on the new housing development site was almost without doubt a renewal of a former landslide. This assessment was matched by the theoretical analysis of the slope. Remedial works involved regrading the slope to a profile that was close to its earlier geometry as well as introducing gravel filled berms. It was not possible to retain the row of houses that had been planned for this area but at least the remaining parts of the development were secured. Details of the technical aspects of this matter are given by Thompson (ref 2).

A claim was made by the housing developer to recover the losses suffered in placing the fill to regrade the slope, partly constructing the houses and then demolishing them, and finally removing fill and forming a berm. The party involved in giving independent advice was asked to provide expert opinion. Documents were disclosed, meetings between experts were held but the issue of whether the problem could have been foreseen and whether the original adviser had acted with reasonable skill and care remained. The legal representatives duly made preparations for a High Court hearing. In the event a settlement was reached on the steps moments before the parties entered the courtroom. Considerable costs were incurred but the potential for saving some monies by not proceeding with the hearing was achieved. It demonstrated the value of reaching a resolution even at a relatively late stage.

Highways Project in Central England (5)
During the construction across challenging terrain of a project involving a wide range of geotechnical aspects, difficulties were experienced which led to cost increases. Each problem was carefully examined as the works proceeded and in places revised or modified designs were introduced. Some matters affected the proposed embankments while others impacted on cuttings. There was a direct and an indirect influence on earthworks volumes. A major retaining structure came under close scrutiny while the foundations to some other features appeared to not necessarily proceed according to plan. The overall result was a marked increase in the final cost for the works compared with the original tender amount.

An expert was appointed to advise the client on whether the difficulties could have been foreseen. Some possible shortcomings were identified and the client entered into discussions with the parties concerned in an endeavour to resolve matters at this stage. No agreement was reached and therefore the client elected to move towards a hearing in an endeavour to recover some of the additional monies expended in the construction of the works. Inevitably on a major construction project, there are considerable volumes of documents held by the various parties. These had to be reviewed by the technical experts and the legal representatives. Meetings between experts were held to refine the issues and, where possible, narrow the differences.

Dialogue between the parties recognised the major cost of embarking on a hearing and therefore it was decided prudently that the overall matter should be divided into a number of sub-trials, each examining a different technical aspect of the works. After the first sub-trial was complete the parties agreed to a settlement of all the matters. Although substantial legal and technical expert costs were involved, as well as the internal costs of the organisations concerned, some marked savings were made by not holding all the planned sub-trials. The example shows the value still to be gained by parties maintaining a dialogue even at the late stage while a hearing is underway and striving to obtain a resolution.

Subsidence in West Midlands of England (6)

Shortly after houses were completed on a new development, a number of them began to suffer undue total and differential settlement. The greatest tilt recorded was in excess of 200mm. The houses were located on a backfilled opencast coal mining pit with the depth of backfill being about 10m. Shallow pillar and stall type mine workings underlay the opencast pit. The backfill had been placed without controlled compaction. Before the houses were built it was decided to inject grout to stabilise the shallow mine workings area under and immediately adjacent to the properties. Stone columns were introduced into the fill by the vibrocompaction process in an endeavour to improve the bearing capacity and reduce the settlement that might otherwise arise as a result of the uncompacted nature of the fill. A performance specification was to be met. Figure 5 shows a typical cross-section of the site.

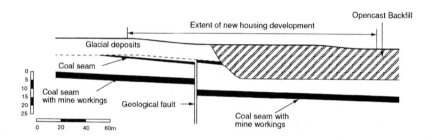

Figure 5. Cross section through backfilled opencast coal pit in West Midlands

The grouting works resulted in the injection of a reasonable but not excessive amount of grout and gave no reason for concern that features such as undetected shafts might exist. The stone columns were partially penetrating with respect to the full depth of the fill. Indeed some were of only modest length. When the settlements occurred, an independent expert was appointed to review the matter, advise on whether one or both of the ground improvement processes might be defective and give guidance on possible remedial works. The first stage was to conduct an investigation of the fill and the underlying solid strata. It was established that the fill consisted of loosely compacted lumps of stiff clay. The depth to the most shallow coal seams was confirmed together with an indication of whether any open voids might still remain. With respect to the fill it was recognised that the stone columns could be acting as vertical drains introducing water into the fill thereby giving rise to inundation collapse. Concerning the shallow mine workings, it was decided to regrout these as part of a remedial works programme. In so doing it was established that the additional amount and pattern of the further grout takes bore no comparison with the pattern of settlement. Accordingly it was deemed very unlikely that inadequate grouting was the cause of the problem. However the parties concerned could not reach a settlement at this stage.

The documents held by the parties were duly examined, reports written and meetings between experts were held. At the same time the legal representatives drew up arrangements for a High Court hearing. Because of the considerable complexity of the matter no early settlement was achieved and the judge recognised that a hearing was largely inevitable if justice were to be done. The trial lasted about four months with the judgement being given some months later. It was stated that it was a case that should have been settled at an early stage. In the absence of such a settlement the costs of the hearing were broadly comparable with the amount of damages. This final case history shows that even though a staged approach may be followed, the occasion arises when all potential for saving costs by reaching an early settlement is extinguished.

DISCUSSION

In a document entitled Creating value in Engineering (ref 3) a diagram was produced which divided the engineering processes of a project into various stages from conception through to completion and on to operation. It demonstrated that early input of engineering advice gave the greatest opportunity for maximising potential cost savings. If such an input were deferred until problems arose then not only was much of this potential lost but there were also significant costs incurred in introducing changes necessary to achieve a successful outcome. This approach of added value was discussed by Thompson (ref 4). The concept can be directly applied to the resolution of engineering claims.

The first two case histories above achieved an early settlement, one being before any litigation was entertained. Although engineering changes were inevitable, most if not all the potential costs associated with legal and expert input were saved. These are plotted as points 1 and 2 on figure 6. In the third and fourth case histories, the stage of disclosure of documents was reached. Expenditure was incurred in examining these, preparing expert reports and holding meetings between experts and of the legal representatives. However the rigorous technical expert input allowed many of the issues to be resolved, to the point where settlement was achieved at this stage. There was some loss of potential cost saving but all the expense of a hearing was avoided. These examples are plotted as points 3 and 4 on figure 6.

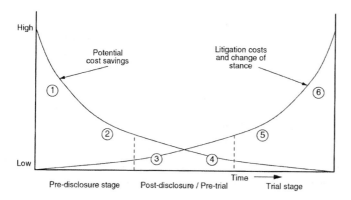

Figure 6. Cost savings and litigation costs versus stage at which settlement reached

The final two examples reached the stage of a hearing in spite of strenuous efforts to identify and narrow the relevant issues.   Each concerned multi-faceted matters of considerable geotechnical engineering complexity and claims of £1million or more.   Although much of the potential for cost saving was lost by this stage, some still remained.   In case history 5 the agreement between the parties to adopt various sub-trials and the settlement after the first of these at least achieved some savings.   The last example shows how all potential for financial savings is destroyed, with attention being drawn to this loss in the judgement.   These case histories 5 and 6 are plotted in their respective positions on figure 6.

The early introduction of expert technical input when claims arise is analogous to the opportunity to change engineering designs by early technical input in a project.   An expert can advise a claimant or a defendant that his position is not as strong as first thought.   Often only a slight change in the stance of one or both sides can achieve a resolution.   This is comparable to the modest cost to change engineering designs at the concept stage.   When a matter progresses past this stage, detailed forensic engineering examination assists in the full and proper identification of the issues.   Again the sides may decide to resolve their differences but they are aware that monies have been spent on both legal and expert input.   They may seek to recover these and their position becomes more entrenched.   A more marked change of stance is then required to achieve a settlement.   A matter can almost gain its own momentum with increased polarisation of the parties.   Once a trial commences there is a yet greater extent of change of position necessary to avoid the trial proceeding to the point of judgement.   This increase in change of stance and the increased litigation costs as a matter proceeds towards and into trial is depicted on figure 6.   It demonstrates that, at the same time, there is a reduction in the potential for cost savings.   The early introduction of a technical expert and the adopting of a staged approach to the forensic examination of the relevant information can allow parties to recognise that an initial modest change of stance may achieve a resolution. The cost savings are maximised.   Such an approach accords with the broad way forward encouraged by Lord Woolf and parties should endeavour to follow it from the outset.

ACKNOWLEDGEMENTS
The author would like to thank Beachcroft Wansbroughs, solicitors; Gateley Wareing, solicitors; Watson Burton, solicitors; The Highways Agency and Collegiate Management Services for granting permission to refer to some of the cases above.   In addition permission was given by a housing developer and various consulting engineers in respect of some of the matters to which reference is made and is duly appreciated by the author.

REFERENCES
1. Leach B.A. and Thompson R.P. The influence of underdrainage on the stability of relict landslides. 9th European Conf. on Soil Mechanics and Foundation Engineering. Dublin 1987.
2. Thompson R.P.   Stabilization of a landslide on Etruria Marl.   Conference on Slope Stability Engineering. Isle of Wight. Institution of Civil Engineers. Thomas Telford 1991.
3. Creating value in Engineering. Design and Practice guide. Institution of Civil Engineers. Thomas Telford 1996.
4. Thompson R.P. The value of timely hazard identification. Seminar on The Value of Geotechnics in Construction. Institution of Civil Engineers 1998.

Investigation of failures of marble facades on buildings

IAN R. CHIN
Vice President and Principal
Wiss, Janney, Elstner Associates, Inc.
Chicago, Illinois USA

INTRODUCTION

Prior to the 1960s, the structural design of load bearing and non-load bearing walls on buildings was primarily based upon empirical design principles. According to an early building code (1), this design approach, when applied to a 100 ft (30m) tall building resulted in walls that vary in thickness from 16 in. (406mm) at the top to 24 in. (610mm) at the bottom of the wall for load bearing walls; and walls that were constantly 12 in. (305mm) thick over the full height of the building for non-load bearing walls. The empirical design requirements of this code also required the minimum thickness of stone ashlars for load bearing and non-load bearing walls to be 4 in. (100mm).

Due to the effect of the weight of stone in exterior walls on increasing the construction cost of buildings, designers in the 1960s began to reduce the thickness of stone veneers on high rise and low rise buildings through the use of rational design principles to the point where the thickness of stone veneers on buildings was reduced to be typically 1-1/4 in. (30mm) thick for solid stone panels and 7/8 in. (20mm) thick for veneers on stone faced precast concrete panels.

The rational design principles used by designers to determine the thickness of stone veneers on buildings typically includes a structural analysis of the stone panel and its connection to the building that utilizes the design wind load, the span of the veneer or panel between connection points, the initial flexural strength of the stone, the effects of weathering on the flexural strength of the stone, and a design safety factor for the stone. When this design approach includes all of these design parameters, the design of the stone panels is usually adequate. However, in numerous instances where design parameters such as stresses in the stone at connection points and/or the effect of weathering on the flexural strength of the stone were not included in the rational design approach, failure of the stone panels have occurred to the extent that removal of all of the stone panels from the building is necessary to insure the safety of the occupants and of people around the building.

PERFORMANCE OF MARBLE PANELS ON BUILDINGS

The evaluation of marble panels on several buildings by the author and his co-workers has revealed that, generally, marble panels with a minimum thickness of approximately 4 in. (100mm) are durable and have performed well even on buildings that are over 50 years old. An example of a 40 year old building in Chicago, Illinois, USA that is clad with 4 in. (100mm) thick marble panels that are performing well is shown in Figure 1.

Forensic engineering: the investigation of failures. Thomas Telford, London, 2001.

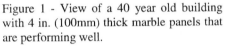

Figure 1 - View of a 40 year old building with 4 in. (100mm) thick marble panels that are performing well.

Figure 2. View of a building with 1-1, (30mm) thick marble panels that had to l removed about 18 years after the building w constructed.

The evaluation also revealed that thin marble panels with a maximum thickness of approximately 1-1/4 in. (30mm), are generally not durable and have deteriorated significantly within about 16 years of their installation on the building. Figure 2 is a view of the 82 story tall Amoco Building in Chicago, Illinois, USA, that was clad with 1-1/4 in. (30mm) thick marble panels that had to be removed due to significant distress in the marble panels that was caused by a significant reduction in strength in the marble due to exposure to the weather. The distress primarily consisted of extensive cracking of the marble panels at and away from the panel connections.

INVESTIGATION OF DISTRESSED MARBLE PANELS ON BUILDINGS

The author and his co-workers have investigated distressed marble panels on several major buildings in the United States. These buildings are located in cities in seven different states within the United States and at latitudes from approximately 35 to 43 degrees north. The marble on the buildings investigated include marble from Carrara, Italy and from the State of Georgia in the United States. The thickness of the marble panels investigated vary from 7/8 in. (20mm) up to 1-5/16 in. (33mm).

The investigation of the marble panels on these buildings by the author and his co-workers generally included:

1. Hands on, close-up visual inspection and documentation of the condition of the marble panels.
2. Laboratory testing of samples of marble panels removed from the building.

3. In-situ structural load testing of marble panels on the building.
4. Structural analysis of the marble panels.

During these investigations, marble panels were removed from the exterior façade of the buildings and tested in the laboratory in accordance with the American Society for Testing and Materials (ASTM) to determine their flexural strength per ASTM C880 (2), and to check conformance of the marble with ASTM C503, "Standard Specification for Marble Dimension Stone (Exterior)" (3). When available, attic stock or replacement panels that were stored in a temperature controlled environment in the building were also obtained and were tested to determine the original properties of the marble at the time the panels were installed on the building. For some of the buildings investigated, the marble panels removed from the building were subjected to accelerated weathering testing in the laboratory to obtain an indication of their predicted flexural strength after future years of exposure to natural environment. The accelerated weathering test performed consisted of immersing ASTM C880 stone specimens that were cut from the panels face down to a depth of 1/4 in. (6mm) to 3/8 in. (10mm) in a 4 pH solution of sulfurous acid solution, and subjecting the specimens to cycles of -10 degrees F to +170 degrees F (–23 degrees C to +77 degrees C) at a rate of three cycles per day. Twelve and one half of these cycles have been found to be equivalent to approximately one year of natural weathering in the Chicago, Illinois, USA area (latitude approximately 42 degrees north) (4).

RESULTS OF INVESTIGATIONS

1. **Post-Construction Reduction in Strength of Marble is the Primary Cause of the Failure of Thin Marble Panels**
   The investigation of cracked and/or bowed thin marble panels on the buildings investigated revealed that the failure of the marble panels was primarily caused by a significant reduction in strength of the marble that occurred after the panels were installed on the building, due to exposure of the panels to natural heating and cooling from the weather. The strength reduction caused the marble panels to have an ultimate flexural strength that could not support the design wind loads on the panels without cracking.

   Figure 3 is a graph of ASTM C880 flexural strength vs years of exposure of marble panels investigated on 11 major buildings in the United States. This data revealed the following:

   • The marble panels on all of the 11 buildings investigated underwent significant reduction in flexural strength due to exposure to normal weathering after they were installed on the buildings. The reduction of flexural strength varies from about 25 percent to 45 percent after 10 years of exposure, and from about 35 percent to 70 percent after 30 years of exposure to the weather.
   • The reduction in flexural strength of the marble panels due to exposure to the weather continues indefinitely at a reduced rate. Refer to building nos. 2, 4, 6, and 11 in Figure 3.

- The rate and extent of the reduction in the flexural strength of marble from Carrera, Italy, and of marble from the State of Georgia, USA, are similar. Refer to building nos. 6 and 11 in Figure 3.
- Thinner marble panels lose more flexural strength than thicker marble panels over the same period of time and exposure conditions. Refer to building nos. 5 and 9 in Figure 3.
- The initial flexural strength of marble from Carrera, Italy varies from approximately 1,100 psi (7.6 Mpa) to 3,100 psi (21.4 Mpa). Refer to building nos. 1, 2, 4, and 6 in Figure 3.
- Using marble on marble-faced precast concrete panels does not prevent the marble from losing flexural strength. Refer to Building No. 2 in Figure 3.

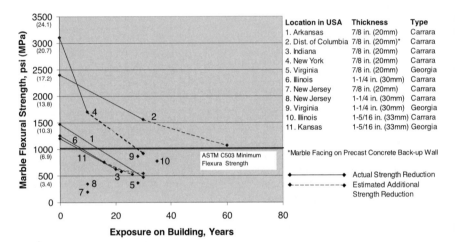

**Fig. 3 ASTM C880 Flexural Strength of Marble
vs. Exposure Time on Building**

2. ***The Loss in Strength that Occurs in Marble is caused by Exposure to Heat and Cold and has been known for Over 125 Years.***
   As chronologically presented below, the propensity for marble to lose strength due to exposure to heating and cooling has been published in marble related publications for over 125 years.

   1919: In the paper entitled "Physical and Chemical Tests on the Commercial Marbles of the United States" that was prepared by D. W. Kessler and published on July 15, 1919, Kessler reported the following (5).

   - The compressive strength of marble samples tested by the Ordinance Department of the United States Army in 1875 was reduced by 53.8 percent after heating at 212 degrees F (100 degrees C) and cooling at 32 degrees F (0 degrees C).

- The compressive strength of "a white calcite marble" tested by Kessler was reduced by 7.3 percent after being heated 100 times to 300 degrees F (150 degrees C).
- The compressive strength of a "a white calcite marble" tested by Kessler was reduced by 7.4 percent after "30 freezings".

1932: In the report entitled "The Weathering of Natural Building Stones" prepared by R. J. Schaffer (6), Schaffer reported the results of the above Kessler tests on reduction in compressive strength of marble after exposure to heated and cooling.

1958: In Building Construction Handbook, Frederick S. Merritt, editor (7), also reported the results of the above Kessler tests on reduction in compressive strength of marble after exposure to heating and cooling.

1962: The Marble Institute of America, Inc. stated the following in their Marble Engineering Handbook (8):

- "Experimental evidence from Armour tests indicates that continuous heating and cooling of marble causes some slight residual expansion due to loosening of the crystal bonds. This heating and cooling diminishes the strength of marble, which depends on the strength of the crystal bonds".
- "Laboratory tests at Armour Research Foundation show the flexural strength of marble to be reduced by aging. In order to compensate for the aging effect, the ultimate strength of marble was considered as half of its actual tested strength".
- "The criterion for calculating allowable stresses in the design of transversely-loaded marble walls is to multiply one-half the ultimate failure stress by 40 percent. This will provide a total safety factor of 5, which the Marble Institute recommends for contemporary curtain wall and veneer design".

1966: In Time-Savers Standards, (9) Arthur Hockman of the National Bureau of Standards, U.S. Department of Commerce reported the above published statements by the Marble Institute of America.

1976, 1983, 1987, 1991, and 1999: The Marble Institute of America in their publications (9) for these years continued to recommend a factor of safety of 5 for the design of exterior marble walls and veneers as originally recommended in their 1962 Marble Engineering Handbook and which was based upon tests that demonstrated that "heating and cooling diminishes the strength of marble".

3. ***Marble Expands Permanently when Exposed to Heat. This Property of Marble has been known for over 100 years.***
As chronologically presented below, the propensity for marble to undergo permanent expansion when heated, that is to expand when exposed to heat and not to return to its original length when it is cooled to its original temperature (hysteresis), has been publicized in marble related publications for over 100 years.

In his paper, D W. Kessler (5) reported the following results of measurements made of marble specimens by other researchers:

1890: A series of measurements made by William Hallock of the United States Geological survey of ten, 3 ft (1m) long marble specimens at temperatures of 68 degrees (20 degrees C) and 212 degrees F (100 degrees C) revealed that the specimens had a permanent increase in length of 1/125 to 1/85 in. (0.2 to 0.3 mm).

1910: Measurements made by N.E. Wheeler on a 1 in. (24 mm) diameter by 8 in. (200mm) long Carrara marble specimen at six different sets of temperatures up to approximately 930 degrees F (500 degrees C) revealed that the specimen had a permanent increase in length of 1/24 in. (1.095 mm).

1917: Measurements made by L. W. Schad and P. Hidnert of a 3/8 in. (10mm) square by 12 in. (300 mm) long "Pittsford Italian" marble from Vermont after three heatings between –13 degrees F (–25 degrees C) and 500 degrees F (300 degrees C) revealed that the specimen had a permanent increase in length of 1/25 in. (0.96 mm). Measurements made by Schad and Hidnert on a similarly sized specimen of "Florentine" marble from Vermont after five heatings from 68 degrees F (20 degrees C) up to 570 degrees F (300 degrees C) revealed that the specimen had a permanent increase in length of 1/25 in. (1.04 mm).

1919: In his paper, D. W. Kessler (5) reported that his measurements of a marble sample at temperatures of 32 degrees F to 140 degrees F (0 degrees C to 60 degrees C) revealed that "the sample did not regain its former length, but remained slightly longer".

1932: In his report entitled "The Weathering of Natural Building Stone," R. J. Schaffer (6) reported "Kessler showed that, after heating, marble does not contract to its original length, but suffers a permanent set".

1958: In Building Construction Handbook, Frederick S. Merrit (7) reported that "Wheeler measured the permanent expansion that occurs on repeatedly heating marble".

1962: In their Marble Engineering Handbook, the Marble Institute of America, Inc. (8) stated the following on the hysterisis of marble: "laboratory tests for the coefficient of thermal expansion of marble indicates that after several cycles of heating and cooling, a residual expansion of about 20 percent of the original calculated thermal expansion can be expected".

1976, 1983, 1987, 1991, and 1999: In their publications for these years, the Marble Institute of America, Inc. (10) stated "laboratory tests for the coefficient of thermal expansion of marble indicate that after several cycles of heating and cooling, a residual expansion of about 20 percent of the original increase can be expected." In addition, the 1991 and 1999 editions of the Marble Institute of America's Dimension Stone state: "Hysteresis is a phenomenon that affects certain "true" marbles. Unlike most stones which return to their original volumes after exposure to higher or lower temperatures, these marbles show small increases in volume after each rise in temperature above the starting point".

4. **The permanent expansion of marble has caused bowing of marble panels on buildings**
Outward and inward bowing of marble panels up to approximately 1-1/4 in. (30mm) has occurred on several of the buildings investigated by the author and his co-workers.

Bowing of the extent observed in 1-1/4 in. (30 mm) thick marble panels has not been observed in thick 4-in. (100mm) thick marble panels.

The observed bowing of marble panels on the buildings investigated has usually been most extensive and pronounced on the south, east, and west elevations of the building which are exposed to the direct rays of the sun. When bowing of marble panels occurs on the north elevation of a building, the bowing usually occurs at a significantly lesser degree. This observed pattern of bowing confirms that the bowing is related to the amount of heat the panels are exposed to, and strongly indicates that outward bowing of the panels was caused by differential permanent thermal expansion between the exterior surface region of the panels that is directly exposed to the sun and the interior surface region of the panels that is not directly exposed to the sun. The Marble Institute of America in their 1991 and 1999 publications of <u>Dimension Stone</u> (10) states this differential expansion within the stone panel is more likely "accommodated or restrained in thick veneers than in thin ones," and that "if it is not restrained, bowing of the marble panel ensues". Kessler has also reported (5) on the upward bowing of 2 to 3 in. (50 to 76mm) thick horizontal marble slabs found in a cemetery in Havana, Cuba. Kessler concluded that the parts or layers of the marble panel retaining the greater permanent increase would "outgrow" the other parts by warping.

Marble panels on marble-faced precast concrete panels on buildings investigated had insignificant bowing due to the back of each panel being connected to the concrete with multiple metal connectors.

The cause(s) of the inward bowing of marble panels on buildings have not yet been fully evaluated.

5. ***Marble Panels with the Largest Bow have the Lowest Flexural Strength***
    During the investigation of distressed marble panels on buildings by the author and his co-workers, marble panels with varying degrees of outward bow, that were not backed up with precast concrete, were removed from the façade on some of the buildings and were tested in the laboratory. The results of the testing are shown in Figure 4. The testing revealed that, in general, the marble panels with the largest bow have the lowest flexural strength and that the panels with the smallest bow have the highest flexural strength. The test results, therefore, revealed that the marble panels with the largest bow have experienced the largest reduction in flexural strength.

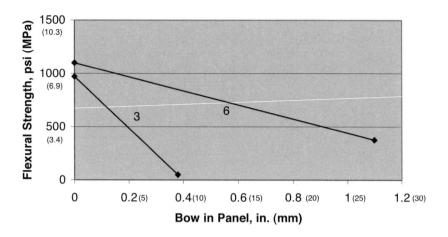

**Fig. 4 ASTM C880 Flexural Strength of Marble
vs. Extent of Bow in Panels**

CONCLUSIONS

The investigation and evaluation of distressed marble panels on several buildings by the author and his co-workers have revealed the following:

1. Marble panels with a minimum thickness of approximately 4 in. (100mm) are durable and have performed well even on buildings that are over 50 years old.
2. Since the 1960s, thin marble panels that are approximately 1-1/4 in. (30mm) thick have been used on several buildings. These thin marble panels have proved not to be durable. On buildings investigated by the author and his co-workers, thin marble panels developed significant cracks and/or bowing within about 15 years of their installation on the building and have had to be removed from the building to ensure the safety of the occupants and of people around the building.
3. The cracking of the thin marbles panels was caused by a reduction in flexural strength of the marble of up to 70 percent due to exposure of the panels to natural heat and cold from the weather.
4. The reduction in flexural strength of marble continues indefinitely at a reduced rate.
5. The rate and extent of the reduction in flexural strength of marble from Carrara, Italy and of marble from the State of Georgia, USA are similar.
6. Thinner marble panels lose more flexural strength than thicker marble panels during the same time period and exposure.
7. Using marble on marble-faced precast concrete panels does not prevent the marble from losing flexural strength.
8. The propensity for marble to loose strength due to exposure to heat and cold has been known and published in marble related publications for over 125 years.
9. Marble expands permanently when exposed to heat (hysteresis). This property of marble has been known and published in marble related publications for over 100 years.

10. Permanent expansion of marble has occurred in thin marble panels that are directly exposed to the sun. The differential permanent expansion that occurs between the outside and inside faces of marble panels has caused thin marble panels to bow outward.
11. In general, marble panels with the largest bow have the lowest flexural strength and, therefore, have experienced the largest reduction in flexural strength.

RECOMMENDATIONS

The following design steps will help to minimize the development of cracks and bowing in marble panels on buildings due to loss in strength and permanent expansion of the marble from exposure to the weather:

1. Determine the amount of strength loss that will occur in the specific marble that is selected for the building. This determination can be made by testing ASTM C880 specimens of the marble removed from representative blocks of marble mined specifically for the building. The ASTM C 880 test should be made prior to, during, and after the performance of the laboratory accelerated weathering test presented in this paper.
2. Use the test results from item 1 in a structural engineering analysis to determine the thickness, size, and connections of the panels to the building structure.
3. For preliminary design of the panels, historical ASTM C 880 test results may be used in the analysis. However, since marble is a product of nature and its properties can vary significantly within and between each quarry, testing of marble samples representing the actual marble that will be used on the building as described in item 1 above and in accordance with ASTM C 503 (10) must be performed to finalize and complete the design.
4. Design expansion joints around each marble panel to accommodate the expected permanent expansion of the marble and other building movements.
5. Use marble faced precast concrete panel system or reasonably sized individual marble panels that are at least 4 in. (100mm) thick to reduce bowing of the panels.

REFERENCES

1. Building Code of the City of New York, 1927, pp 62-65, Herald Square Press, New York, NY, USA.
2. American Society for Testing and Materials, ASTM C 880 "Standard Test Method for Flexural Strength of Dimension Stone," Annual Book of ASTM Standards, Vol. 04.05, Philadelphia, PA, USA.
3. American Society for Testing and Materials, ASTM C 503, "Standard Specification for Marble Dimension Stone (Exterior)," Annual Book of ASTM Standards, Vol. 04.05, Philadelphia, PA, USA.
4. Bortz, S.A. and Wonneberger, B., "Laboratory Evaluation of Building Stone Weathering," in Degradation of Natural Building Stone, American Society of Civil Engineers, 1801 Alexander Bell Drive, Reston, VA, USA.
5. Kessler, D.W., "Physical and Chemical Tests on the Commercial Marbles of the United States," in Technological Papers of the Bureau of Standards, July 15, 1919, pp. 13, 17, 28, and 40, Superintendent of Documents, Government Printing Office, Washington, D.C., USA.
6. Schaffer, R. J., "The Weathering of Natural Building Stones," Department of Scientific and Industrial Research, Building Research, Special Report No. 18, 1932, p. 43, His Majesty's Stationary Office, London.
7. Merritt, F.S., Building Construction Handbook, 1958, pp. 2-25, McGraw-Hill, Inc. New York, NY, USA.

8. Marble Engineering Handbook, 1962, pp. 12, 14, and 19, Marble Institute of America, Inc., 32 South Fifth Avenue, Mount Vernon, NY, USA.

9. Hockman, A., "Physical Properties of 113 Domestic Marbles," in Time-Saver Standards, Fourth Edition Callender, J.H., Editor-in-Chief, pp. 432, McGraw-Hill, Inc., New York, NY, USA.

10. Marble Design Manual, 1976, pp. 0004.04; Design Manual II, 1983, pp. 0004.04; Dimension Stone – Volume III, 1987, pp. 0004.04; Dimension Stone-Volume IV, 1991, pp. M-14; Dimension Stone-Volume VI, 1999, pp. 48; Marble Institute of America, Inc. 33505 State Street, Farmington, MI 48335, USA.

Material Evidence:
the forensic investigation of construction and other geomaterials for civil and criminal cases

IAN SIMS
Director, Materials Consultancy, STATS Limited, St Albans, UK

ABSTRACT

Forensic investigations into construction failures are considered that have arisen variously from the nature, compliance, fitness, compatibility, performance and application of materials. Every case tends to be different, but some detailed flaws in materials, products or practice are recurrent and avoidable. It is shown, by the presentation of examples from actual experience, that most of the costly post-construction disputes were predictable and could thus have been prevented by the procurement of independent specialist advice at the time of construction. Cases are selected from a range of materials and divided into 'unsuitable materials' and 'unsuccessful application', though in reality most failures occur for a combination of reasons.

INTRODUCTION

Disputes over the causes of failure in buildings and civil engineering structures frequently include the need to investigate the nature, compliance, fitness, compatibility and performance of materials, or the way in which materials have been used. Every technical dispute in building technology or civil engineering is different, so that engineering experts frequently have to work from first principles to augment their experience. Nevertheless, there are recurring faults and flaws, sometimes in connection with particular materials or products, at other times related to inappropriate practice and even, occasionally, caused by inadequate or actually misleading specification or guidance. Materials-related failures can frequently be attributed to foreseeable and avoidable circumstances, demonstrating that costly and disruptive post-construction disputes could usually have been prevented by the procurement of specialist advice, independent of suppliers or contractors, at an early stage in the project.

The following sections seek to illustrate the causal link between inadequate or misleading advice during construction and the later development of time-consuming and expensive disputes. These examples are taken from selected cases over a 25-year period of involvement in which the author has acted as an expert witness, some of which were eventually the subject of a settlement whilst others actually reached the courtroom. It is of course necessary for the identity of these cases to remain anonymous.

UNSUITABLE MATERIALS

Investigations into engineering defects or failures rarely identify simple singular causes. It is much more common for causal mechanisms to involve a combination of factors. Consequent

Forensic engineering: the investigation of failures. Thomas Telford, London, 2001.

litigation typically develops into an inter-related series of issues. The following separation, for the purposes of illustration, into 'unsuitable materials' and 'unsuccessful application' is thus usually an over-simplification.

Rock & Aggregates

Civil engineering and building uses large quantities of particulate rock, variously for rock fill (Smith 1999) and as aggregates in unbound, bitumen-bound and cement-bound materials (Smith & Collis 2001). Mostly, such materials are inert and stable, but occasionally they are potentially unsound and can lead to failure. Such unsoundness is identifiable in advance of use, from informed and/or experienced materials assessment.

Hearting fill to a marine causeway was required to be crushed rock and, in the Specification, the engineer had unwisely listed rock types considered to be suitable, including 'sandstone'. A local material was undeniably 'sandstone' by any geological classification, but was quite unsuitable for the proposed purpose and very different from the 'sandstone' with which the engineer was evidently familiar. The sandstone used was reasonably competent in its dry state, but had a cementing matrix of clay and iron oxide that became very unstable on wetting and was rapidly scoured by water flowing through the fill after placement. This illustrates both the danger of specifying by rock name and the value of petrographic assessment.

A major local authority had the need to establish a replacement source of surface dressing aggregate for their road network, following many decades of success with a crushed igneous rock from a quarry that had closed. The task was entrusted to a firm of Consulting Engineers that identifed a range of existing and potential quarries, sub-contracted to nearby laboratories a programme of British Standard tests mainly for mechanical properties, and eventually enthusiastically endorsed a trachyte source for development as a new road aggregate quarry. The qualities were said to exceed those of the previously used material. There had been no geological appraisal of the site and the petrographical examinations commissioned for selected samples had been used only to cite the correct petrological rock name.

After just one winter of service, road surfaces re-dressed with the new aggregate material were exhibiting widespread failures. Aggregate particles variously fragmented, disintegrated or debonded from the bitumen binder. Subsequent geological investigation found that the rock body was heavily sheared, secondarily mineralised and altered, all of which was easily discernible from natural outcrops that had existed prior to the commencement of quarrying. Moreover, at a microscopic level, the rock was typically micro-brecciated and secondarily mineralised; observations that had been recorded in the earlier petrographic descriptions, but the significance not appreciated by the investigating engineers. New testing by the magnesium sulfate soundness method, which was not at the time a BS test but nevertheless in common use by materials specialists, clearly revealed the unsound nature of the material. It was also found that problems with bitumen affinity had been encountered in tests during the earlier assessment, as might have been predicted from the petrography, but these had been discounted after several repeats had finally obtained acceptable results.

This important example illustrates the need for geological materials to be assessed by scientists with appropriate expertise and experience. All of the problems that were subsequently discovered in service, to the considerable cost of the council tax payers, could and should have been identified in advance by apposite field and laboratory appraisal.

A third example concerns sand that gave rise to alkali-silica reaction (ASR) in concrete. ASR can be expansive and occasionally very damaging to concrete structures. It occurs as the result of reaction between alkali hydroxides mainly derived from cement and certain types of silica that can be present within some aggregate constituents (Swamy 1992, BRE 1999). The intensity of reaction and the magnitude of any resultant expansion is controlled in a complicated manner by several factors. In the case in question, which involved a large public complex of structures, ASR had occurred within a concrete mix containing high-alkali cement, a non-reactive crushed limestone coarse aggregate and marine sand with a potentially reactive component (chert). That some ASR had occurred was common ground, but the seriousness of the resultant damage was disputed. This example concerns an ancillary issue.

It is frequently argued that materials suitability, or 'fitness for purpose', can be assessed with reference to the appropriate BS specification, if such exists. In reality, not all standard specifications are of equal worth in this regard, but it can nonetheless prove difficult to sustain a case of unsuitability for a material that complies in all respects with an appropriate standard specification. In the case in question, the sand used would have needed to comply with the long-superseded version of BS 882 (1965) that was current at the time of construction. This now obsolescent standard included some specific requirements that could be demonstrated by testing, also to BS methods, but additionally stipulated qualitative requirements that were only defined by extremely generalised wording: *'Aggregates ... shall be hard, durable and clean and shall not contain deleterious materials in such a form or in sufficient quantity to affect adversely the strength at any age or the durability of the concrete'*. Examples were given, including clay, flaky or elongated particles, mica, shale, coal and other organic impurities, pyrite and soluble sulfate salts. These did not include potentially alkali-reactive constituents. Indeed, in 1965, it was widely believed that ASR was not a problem encountered with UK aggregates (Sims 1992).

Nevertheless, it was argued in this example case that the sand could not have complied with BS 882, because it had been found to be reactive in practice and thus demonstrably had contained *'deleterious materials in such a form or in sufficient quantity to affect adversely the ... durability of the concrete'*. It was countered that, whilst this might be true in hindsight, it was not a judgement that could or should have been made at the time of construction, because ASR was not then thought by authorities to be a practical problem with UK aggregates and there were contemporary publications that endorsed that view. In other words, the sand had been a suitable material when judged by the state of knowledge that existed at the time.

This consideration applies a critical condition to the role of informed appraisal that was argued above for the cases of unsound sandstone fill and unsound road aggregate. Today, routine petrographical examination of the sand would identify the chert as a potentially reactive constituent and measures could be taken to minimise the risk of ASR. However, had petrography been carried out in the late 1960s, which was not then routine or even common, the chert would doubtless have been identified, but its potential significance not necessarily recognised. Even if a specialist with international experience had appreciated the possible reactivity of chert, any warnings would have been contrary to the contemporary consensus.

Building Stone

Natural stone has long been a traditional construction material and its use in the UK has undergone a major revival in recent years. Stone exhibits both systematic and random variations in composition, texture and weathering condition. Also, in the case of many sources, suitable material can only be extracted from certain locations and with rigorous selection. The traditional indigenous stone industry has increasingly had to compete with an exotic range of imported and competitively-priced materials, many of which have only a limited track record overall and possibly none at all in UK conditions.

York stone is a traditional paving material that continues to be popular. The term 'York stone', however, is a generalised name for a range of thinly bedded sandstones from West Yorkshire that includes materials from various stratigraphic units and a relatively large number of quarry sources. There is no one material that can legitimately be termed 'York stone' and it is important that the properties of particular materials are evaluated.

Several cases have involved York stone paving that has failed prematurely, usually by progressive surface delamination. Commonly these problems have afflicted domestic premises, where the purchaser has innocently placed trust in the skill and honesty of the stone supplier. Failure occurs because the stone supplied is too fissile and/or is already in a weathered state, so that durability has been compromised by natural geological processes. In these circumstances, the integrity of laminations that would be durable in an unweathered stone, have been adversely affected by long-term natural weathering and are thus rapidly exploited by cycles of wetting & drying, and sometimes freezing & thawing, when exposed in a pavement. In some cases, owners have inadvertently worsened the situation by the application of de-icing salts, which magnifies the detrimental effects.

Such unsuitably fissile or weathered York stone material can be avoided by efficient selection procedures at various stages. Geomaterials scientists are available to assist producers, suppliers and contractors in the identification of unsuitable material. Many cases may result from economising on such specialist advice, so that unsuitable material that is not necessarily distinguishable by the unaided eye is unwittingly supplied. In other examples, there have been very clear differences between an unweathered specimen, fortuitously retained by the purchaser, and the weathered material subsequently supplied in bulk.

Limestone is a popular stone for both cladding and paving. Stylolites are unique to limestones and some marbles and are a source of concern to materials scientists owing to their implications for strength performance and durability (Yates & Chakrabarti 1998). These irregular suture-like boundaries form in limestones as the result of pressure-controlled dissolution during diagenesis and undoubtedly form a physical discontinuity within the resultant stone. Most limestones contain stylolites to some degree and clearly they are infrequent causes of bending failures or premature deterioration. However, the inappropriate concurrence of stylolites with positions of stress concentration must always be avoided, also some forms of stylolite can be the cause of premature deterioration.

One case, in the Arabian Gulf, involved a black limestone that was heavily affected by complex stylolites, comprising concentrations of discontinuities and associated clay-like material. The material was used as an interior lining in an air-conditioned shop. Moisture penetration into the stylolitic zones, possibly induced by condensation, caused degradation of the cladding panels with associated efflorescence. The vulnerability of this stone would easily have been identified by petrographical examination.

Roofing slate remains one of the most important applications of natural stone for building purposes. Traditionally, slates used in the UK were derived mainly from North Wales, with smaller production from two or three other regions. However, in recent years, low-cost slate materials of sometimes indifferent quality have been imported from other parts of the world, including Spain and China. Many of these imported slate materials perform satisfactorily, but there have been cases of decidedly premature and very predictable failure.

A number of roofs have exhibited breakage and fragmentation of individual slates, sometimes with associated unsightly brown staining, and these failures have occasionally started to occur before the remainder of the construction work was completed. In some cases, the slate materials appear to comply with BS 680, but fail rapidly in service for reasons that are not expressly covered by the provisions of that specification. Based solely on experience with slates from UK sources, the somewhat obsolescent BS 680 has specific test requirements for three properties that relate to weathering durability: water absorption, cyclic wetting & drying and acid immersion. Even then, the latter is only required for slates that will be exposed in service to a polluted environment. There is no BS test requirement for strength, so that some fragile types of slate can appear to be suitable, as judged by the BS. In one major case, imported slates sampled from a prematurely failing roof were found to satisfy the three testable requirements of BS 680, but to fail the bending strength and more aggressive acid-immersion weathering tests in ASTM C406, the American standard specification.

Premature slate failures are sometimes caused by the presence of tight joints that are not revealed by the traditional slater's 'ringing' method, but which will be susceptible in service to rapid weathering and breakage. Pyrite is a common constituent of slate, but some varieties and/or modes of occurrence are particularly reactive and thus prone to oxidation in service, leading to disruption and disproportionate staining. Fine-grained reactive pyrite sometimes occurs along joint planes, so that expansive oxidation of the pyrite is a principal cause of the opening of joints and resultant breakage. These potentially deleterious features are not proscribed by the specific requirements of BS 680 (another example in which compliance with a BS does not assure suitability), but can be identified by experienced assessment.

Cementitious Products

Concrete blockwork is a common feature of modern building and largely appears to fulfil its function without problems. However, occasionally blocks prove to be unsound and give rise to the cracking of walls or worse, generating disputes.

In the past, blocks made using furnace clinker aggregate were in common use and quite frequently gave rise to unsoundness. The clinker materials were frequently contaminated with unstable constituents, including unburned coal, sulfurous compounds and hard-burned lime (Sims & Brown 1998). Improved practices in the processing of furnace clinker generally eradicated the problems associated with hard-burned lime, so that the current version of BS 1165 only offers controls over unburned coal (by loss-on-ignition testing) and sulfate content. Withdrawal from the BS of measures to detect free lime, which can hydrate and then carbonate in service with resultant expansion, has led to a number of recent cases caused by the production of unsound clinker blocks. Consequential damage ranges from minor surface blemishes on plaster applied to the blockwork, through cracking of the blockwork that is reflected in any attached render of plaster, to structural collapse.

In one notable case, which had caused the collapse of several walls, the block maker had purchased clinker aggregate from a large private factory with its own coal-burning power station. A crucial element of the dispute was that the clinker had been sold as waste and purchased as aggregate. Initial testing of one small sample by a local university had confirmed compliance with BS 1165 (loss-on-ignition and sulfate) and this had been taken as evidence of suitability. Problems did not arise immediately, but only later when the blend of coal supplied to the power station was changed, without effect on its specification as a fuel, and the clinker waste started to contain free lime. The root of this case was the inadequate advice sought by and provided to the block maker at the outset. An informed adviser would have recognised that BS 1165 could not alone provide evidence of suitability with a new source of clinker material and also recommended regular monitoring of the aggregate.

Autoclaved aerated concrete (AAC) blocks are common products in modern building. The main components of AAC are Portland cement, fine-grained furnace ash from modern power stations and abundant air-voids. Some varieties also contain a small proportion of natural sand. These lightweight blocks have a tendency to exhibit greater moisture movement (drying shrinkage and wetting expansion) than dense concrete blocks and this is recognised by a higher maximum limit for drying shrinkage of lightweight blocks in BS 6073 (0.09% instead of 0.06%). In the author's experience the drying shrinkage of AAC blocks is typically very close to this maximum level and, in several disputes arising from the extensive cracking of AAC blockwork, a slightly increased content of residual unburned fuel within the AAC material has seemed to give rise to an unacceptable degree of moisture movement, sometimes two or three times the maximum limit. Mild cases have led to faint cracking of applied plaster that could be repaired, but more severe examples have caused repeated shelling and cracking of the plaster and even instability within the blockwork wall.

These problems are capable of being identified in advance and thus prevented. Ash materials can be monitored for residual fuel content and BS specifications provide limits on loss-on-ignition or even carbon content (BS EN 450). Blocks can be tested for moisture movement characteristics (BS 6073). In the past, for major projects at least, samples from block consignments were routinely subjected to independent compliance tests, but the modern preparedness to rely on manufacturers' own quality systems means that unsound blocks are now only detected through the later investigation of failures.

Fibre-cement products are extensively used for roofing. In the past various types of asbestos fibre were used, but legitimate concerns over health and safety have led to the introduction of non-asbestos fibres, including cellulose and polymers. Roofing products include flat and corrugated sheets, artificial slates and related accessories. In general, experience suggests that the performance of products made with non-asbestos fibres is less reliable than that previously provided by similar products made using asbestos. In particular, several cases involving the splitting of corrugated sheets and the distortion of artificial slates have highlighted an apparent vulnerability to wetting and drying cycles.

Two major and several minor examples have arisen from the failure of corrugated sheets on large factory roofs. Each case was different in detail, but the common factor was the fibre-cement product and its moisture movement characteristics. The dominant defect was the development of multiple longitudinal cracks parallel to the corrugations, initially in the troughs and later also along the crests. Although investigating engineers tended to concentrate on the presence and/or adequacy of movement joints on the roof, each sheet appeared to behave independently. Tests given in BS EN 494 subject specimens to simulated weathering (cycles of soak-dry, freeze-thaw & heat-rain) and monitor any effects by comparing the breaking load before and after simulated weathering. Generally these treatments had a limited effect on breaking load. However, additional monitoring for dimension change indicated predictable movements, notably expansion on wetting and contraction on subsequent drying, but critically also revealed a tendency for gradual net expansion. It was thought possible that the bond between cement and non-asbestos fibres is weaker than that between cement and asbestos fibres, allowing greater materials movement with the former than with the latter. In the case of corrugated products, the overall expansive trend seems to create stresses that effectively seek to 'unfold' the material.

This finding allowed various microscopical observations to be explained. The sheets are manufactured by creating a sheet from the wet mix of cement and fibres, then pressing it into the corrugated form whilst the mix is still plastic. Tension splits or tears might be expected where the plastic sheet is thus stretched, on the top of the crests and the underside of the troughs. Such discontinuous minor tears are a common feature of the product and can be detected in microscope thin-sections later prepared through the sheet thickness, when they typically seem to be 'healed' by secondary deposition. However, after a period in service, the sheets exhibit crack initiation in precisely the opposite positions: on the underside of crests and the top of the troughs. These are interpreted as being generated by the tension arising from the 'unfolding' stresses. As fatigue continues, these cracks extend through the thickness of the sheet, so that visible splits form first along the troughs and later along the crests. It is likely that the restraint of overall movement of the sheet, effected by the individual sheet fixings, will exacerbate the formation of these longitudinal splits.

Such an explanation obviously suggests a generic product defect, as opposed to a deleterious feature that can be identified in advance and then avoided or subjected to corrective or compensatory action. However, not all fibre-cement roofing fails, so that the risk must vary from roof to roof, probably according to a complex combination of factors. Thus, for major projects a programme of preliminary evaluation should identify the materials most at risk of failure in service.

UNSUCCESSFUL APPLICATION

Defective materials are rarely the principal cause of engineering failures. It is much more common for difficulties to arise from the way in which materials have been used, or mis-used, or from an unfortunate combination of circumstances.

Concrete Surface Defects

One large and several ancillary cases involved concrete surface defects on a Caribbean island. The large dispute concerned all the concrete buildings in an extensive beach-side resort and it was alleged that ASR threatened the integrity of the structures, so that wholesale demolition and redevelopment was claimed to be the only remedy. ASR manifested itself as sporadic pop-outs on the soffits of balconies and elevated walkways, also on some retaining walls. There was no cracking or other major damage that was attributable to ASR.

Investigation showed that the aggregate was crushed material from a stratified sequence of volcanic tuffs and sediments. A small proportion of the aggregate particles was potentially reactive, being glassy or chert-like materials. The cement was an imported low-alkali variety and the concrete producer had carefully monitored alkali content in accordance with standard modern practice in the UK. Pop-outs had formed in a certain set of circumstances, where moisture could permeate through the concrete in one direction towards a drying surface. If that surface was coated, for example by the paint that adorned all the concrete within the resort, it would act as a semi-permeable membrane, allowing moisture vapour to egress, but causing transported alkalis to accumulate in the concrete surface zone behind the coating. A local zone high in alkali content was thus created at the concrete surface. Wherever one of the potentially reactive aggregate particles occurred within this alkali-enriched surface zone, highly localised ASR could take place, causing a pop-out (Sims 1996).

It was thus demonstrated that the surface defects were not symptomatic of a more serious occurrence of ASR and that the integrity of the structures were not threatened. As the supplies of both alkali and reactive particles in the concrete zone near the soffit were finite, the impaired surface could be simply repaired and redecorated. A similar occurrence could be avoided in future structures by designing to prevent water permeation through the concrete.

Ageing Cracks in Concrete

Part of a major case concerning structural inadequacy centred on the age of cracking to sub-floor columns. Structural calculations had indicated that the columns had been under-designed for the loadings involved, so that cracking observed on some columns was thought to be evidence of resultant partial failure in service. However, the defendant argued that the cracking was recent damage that had been inflicted by demolition plant, particularly when the loose fill was being removed from around the columns in question.

Core samples drilled through the cracked positions were examined in the laboratory in order to assess the relative age of the cracking. It was thought that evidence that the cracking was long-standing would support the plaintiff's view of over-loading, whereas signs that the cracking was comparatively recent would be consistent with the demolition damage claimed by the defendant. The investigation focused on the degree of carbonation.

All Portland cement concrete gradually carbonates on exposure to the atmosphere (Sims 1994). Cement hydrates, notably calcium hydroxide (portlandite) but also the main hydrates, slowly react with carbon dioxide in the atmosphere to form calcium carbonate. This carbonation proceeds from the exposed outer surface inwards at a generally slow rate that is controlled by a number of factors, including the density of the concrete, the degree of exposure to a constantly replenished source of carbon dioxide, humidity and temperature. In a given set of circumstances, therefore, there is a relationship between the depth of carbonation from the surface inwards and time. If concrete is cracked, the crack acts as a surface breach that facilitates deeper penetration by carbonation along the borders of the crack plane, albeit usually at a slower rate than the general surface carbonation, owing to the limited degree of carbon dioxide replenishment within the confines of the crack opening.

As the concrete was of dense structural quality (the case concerned the number of columns, rather than their quality) and burial in loose fill would have limited exposure to atmospheric carbon dioxide, it was expected that the rate of carbonation would be comparatively slow. Petrographic examination of the concrete in thin-section under a polarising microscope enabled the patterns of complete and partial carbonation to be determined (Sims 1994, St John et al 1998). It was found that significant carbonation penetrated deeply into the concrete along the cracks, suggesting that the cracks were not of recent origin.

Mortar Inadequacy

Mixes for brickwork jointing mortar are usually carefully specified, to ensure compatibility with the masonry units, adequate facility for accommodating movement without inducing damage to the bricks and appropriate weathering durability. Disputes commonly arise out of alleged deviations from the specified mortar mix. Such allegations can be investigated and resolved, one way or the other, by a combination of petrographic examination and appropriate chemical analysis.

The brickwork outer leaf on a very large city centre development was found to be unstable with rapidly deteriorating jointing mortar. A pigmented 1:1:5-6 (by volume) cement:lime: sand mix had been designated, using sulfate-resisting Portland cement apparently in deference to the anticipated high sulfate content of the bricks. A comprehensive programme of mortar sampling was undertaken from many locations around the buildings and at various floor levels, being careful at each location to recover a complete bed of mortar. Microscopical examination of portions in reflected light enabled the cement type to be verified as sulfate-resisting Portland type. Further microscopical examination in thin-section confirmed the constituents as natural quartz/chert sand, Portland cement and iron oxide pigment. It was also clear that the amount of cement binder was decidedly limited overall and, moreover, that even the cement that was present was often concentrated into coarse lumps rather than being uniformly distributed. Chemical analysis confirmed the low cement content, with 95% of samples deviating by 25% or more and 55% having mix proportions as lean as 1:2:12. Thus, although the correct type of cement had been used, its content was much too low and poor mixing had further reduced its effectiveness as a binder.

Floor Surfacing Failure

Floor surfacing failures are quite a common source of dispute. They rarely arise as a result of defects with the surfacing itself, but more frequently because of inadequacies in the subfloor.

The first example concerned many floors within an extensive hospital building. Over time, many depressions ('elephant's footprints') had formed in the vinyl-tiled floor surface, varying in both area and depth. Although the depressions seemed to be randomly distributed on an areal basis, the earliest occurrences could be related to the more heavily trafficked positions, such as lift lobbies and busy corridor junctions. It was alleged by the plaintiff that the screed subfloor was of overall poor quality and would need to be completely replaced.

Investigations were carried out at various locations, wherein comparisons could be made between screed material beneath depressions and that in adjacent positions. The levelling screed was a 'semi-dry' type of cement:sand material with a proprietary additive. Mix proportions were generally close to the specified 1:4, but there was evidence of gross mixing problems, indicated by localised lumps of loose sand and occasional 'balls' of cement. It seemed likely that a free-fall rather than forced-action mixer had been used, although this was denied by the defendant. The surface depressions had formed over the position of sand lumps, probably in immediate response to precise point-loading on the floor, for example from the wheel of a loaded surgical bed or a trolley transporting full gas cylinders. It was thus shown that the depressions were representative of sporadic and potentially repairable localised defects in the screed, rather than indicative of any overall inadequacy.

A second example also concerned a hospital floor, within a kitchen unit. This time a filled-resin proprietary finishing system had debonded from the underlying screed. There were no obvious defects within the semi-dry levelling screed and nothing apparently at fault with the resinous layers. The two materials had simply failed to bond, so that heavy trafficking of the loose and brittle surfacing had caused breakage, followed by ingress of sluicing water into the highly permeable subfloor. The cause was discovered by accident.

Sections were sawn through the resin surfacing and the underlying screed, finely ground, then examined microscopically. Voids in the topmost part of the screed appeared to be infilled with an opaque material of milky coloration. Analysis of this void filling surprisingly found the material to be compositionally indistinguishable from the laboratory resin used to consolidate the screed for sectioning. Further investigation established that the void fillings were indeed the consolidating resin, which had been modified in its visual appearance by reaction with another and incompatible organic substance within the upper part of the screed. As the screed did not contain any organic admixtures, it became apparent that the screed surface had been treated with a material that was incompatible with epoxy resin. The same substance would have been incompatible with the resin surfacing, probably explaining the bond failure. However, no such screed treatment had been specified and it was inconsistent with any of the primers supplied for use with the resin finishing system. It was concluded that a well-meaning but misguided operative had probably 'sealed' the screed surface with an unidentified but apparently alien substance prior to application of the resin surfacing system.

Granite Cladding

Natural stone cladding has become quite fashionable for prestigious and often high-rise buildings. Over a period of experience, it has become recognised that adequate provision must be made in the design of a stone cladding system for the normal and exceptional movements that will take place within each and between individual cladding units. This is largely a matter of designing adequate joints and fixings (Burton 1999). However, the features of the stone must also be taken into account. The problem of stylolites with limestones has been mentioned earlier, but the following example concerns granite cladding.

An important city centre headquarters was clad in panels of light-grey granite. Cracking of the granite was causing concern and the detachment of one corner from a high level naturally gave rise to considerable alarm. It was soon realised that the basic cause of damage to the granite panels was inadequacy in movement provision, leading to the transference of stress from the main concrete structure into the cladding system. Urgent surveys were instigated to quantify the extent of damage and develop a prognosis. These visual surveys, which had to be conducted from window-cleaning gondolas, identified apparently abundant cracking, which were mainly tight and hairline in character, but sometimes widened in the vicinity of fixings and other stress concentration points. Specialised investigation was then undertaken, in order to clarify the nature of these cracks and to identify an objective monitoring method.

Small-diameter cores were drilled from various granite panel locations exhibiting the cracks and these were subjected to petrographic examination in the laboratory. In thin-section, it was clear that many or most of the features mapped as 'cracks' were in fact natural mineral seams within the granite. Cracking of the panels was thus not nearly as advanced as earlier visual inspection had implied, but nevertheless these mineral seams did represent potential planes of weakness when their occurrence coincided with a stressed position. In those cases, any cracking formed preferentially along the seam. A dye-testing technique was adapted and devised for differentiating between mineral seams and fine cracks in granite, including those cracks that had started to develop along mineral seams in critical positions. This dye-testing method is now a regular means of assessing the condition of stone cladding panels, but trained observers are needed to differentiate significant flaws from natural discontinuities including crystal grain boundaries and cleavage planes.

CONCLUSIONS

- Most civil engineering and building disputes involving materials or their usage were predictable and could have been prevented by specialist advice during construction.

- Failures can occur as a result of materials unsuitability, including generic product defects, or the mode of application, but typically there is a combination of causes.

- Standard specifications are found often to be inadequate guides to fitness for purpose.

- Examples of investigations into unsuitable natural materials include unsound rock fill, unsound road aggregate, reactive concrete aggregate and unsound stone or slate.

- Examples of investigations into unsuitable products include unsound concrete blocks and dimensionally unstable fibre-cement roofing materials.

- Examples of investigations into unsuccessful application include alkali-reactivity in concrete, assessing crack age in concrete, brickwork mortar inadequacy, floor surfacing failures and distinguishing real and potential defects in granite cladding.

REFERENCES

ASTM, 1996, *Standard specification for roofing slate*, ASTM C406-89 (reapproved 1996), American Society for Testing and Materials, Philadelphia, USA.

BRE, 1999, *Alkali-silica reaction in concrete*, Parts 1 to 4, Digest 330, Building Research Establishment Limited, Watford, UK.

BSI, 1965, *Specification for aggregates from natural sources for concrete (including granolithic)*, BS 882 & 1201, British Standards Institution, London, UK.

BSI, 1971, *Specification for roofing slates, Part 2, Metric units*, BS 680-1, British Standards Institution, London, UK.

BS1, 1985, *Specification for clinker and furnace bottom ash aggregates for concrete*, BS 1165, British Standards Institution, London, UK.

BSI, 1981, *Precast concrete masonry units, Part 1, Specification for precast concrete masonry units*, BS 6073-1, British Standards Institution, London, UK.

BSI, 1994, *Fibre-cement profiled sheets and fittings for roofing - product specification and test methods*, BS EN 494, British Standards Institution, London, UK.

BSI, 1995, *Fly ash for concrete - definitions, requirements and quality control*, BS EN 450, British Standards Institution, London, UK.

Burton, M (ed), 1999, *Designing with stone*, Ealing Publications Ltd, Maidenhead, UK.

St John, D A, Poole A B, Sims, I, 1998, *Concrete petrography - a handbook of investigative techniques*, Arnold (now Butterworth-Heinemann), London, UK.

Sims, I, 1994, The assessment of concrete for carbonation, *Concrete*, 28 (6) 33-38.

Sims, I, 1992, Alkali-silica reaction - UK experience, Chapter 5, 122-187, in: Swamy, R N (ed), *The alkali-silica reaction in concrete*, Blackie & Son Ltd, Glasgow, UK.

Sims, I, 1994, The assessment of concrete for carbonation, *Concrete*, 28 (6) 33-38.

Sims, I, 1996, Phantom, opportunistic, historical and real AAR - Getting diagnosis right. In: Shayan, A (Ed), *Alkali-Aggregate Reaction in Concrete*, Proceedings of the 10th International Conference, Melbourne, Australia, August 1996, 175-182.

Sims, I, Brown B V, 1998, Concrete aggregates, Chapter 16, 903-1011, in: Hewlett, P C (ed), *Lea's chemistry of cement and concrete*, 4th edition, Arnold, London, UK.

Smith, M R (ed), 1999, *Stone: building stone, rock fill and armourstone in construction*, Engineering Geology Special Publication No 16, Geological Society, London, UK.

Smith, M R, Collis, L (eds), 2001, *Aggregates - sand, gravel and crushed rock aggregates for construction purposes*, 3rd edition (revised by Fookes, P G, Lay, J, Sims, I, Smith, M R & West G), Engineering Geology Special Publication No 17, Geological Society, London, UK.

Swamy, R N (ed), 1992, *The alkali-silica reaction in concrete*, Blackie & Son Ltd, Glasgow, UK.

Yates & Chakrabarti, 1998, *Stone cladding panels: in-situ weathering*, Information Paper IP 18/98, Building Research Establishment Limited, Watford, UK.

The assessment of mechanical failures

M.J. NEALE. FREng FIMechE

INTRODUCTION
This paper discusses the differences and similarities between mechanical and civil engineering failures. A feature of the reliability of mechanical equipment is that it relates closely to the quality of the equipment purchased, as well as to the resources directed to its maintenance. The mechanical equipment used in civil engineering installations tends to look robust and therefore able to accept cost cutting by reduced maintenance activity. However after a few years of reduced maintenance, disaster can set in, and the cost of recovery is usually many times the initial savings.

MECHANICAL ENGINEERING FAILURES
Many mechanical engineering failures tend to differ from civil and structural engineering failures, in that they occur at relatively moving surfaces. This is because it is at these parts of machines that there is the maximum energy concentration in terms of the combination of load and sliding speed. The components which fail most frequently are therefore bearings, gears, piston rings, couplings and seals or any other machine parts with these characteristics. The investigation of mechanical engineering failures, therefore requires a good working knowledge of these components. If such a component does fail, it does not mean that it is defective, because it may be just that it is in a position that makes it sensitive to system effects.

For example, if there is a system consisting of a turbine driving a pump via a gear box, and the teeth break off one of the gears, it may be due to a defective gear. However, equally it may have arisen because the inertias of the two rotors and the torsional elasticity of the system coupling them, results in a torsional resonance close to the running speed, or to the running speed multiplied by the number of blades on the pump rotor, which can give rise to torque pulses at this frequency.

MULTIPLE CAUSES
Another factor that needs to be considered, particularly in forensic work, is that the investigator is often asked to discover the cause of failure. In practice it is extremely rare for a failure to have a single cause. The common situation is that a number of factors occur simultaneously or sequentially, and the critical combination gives rise to the incident. In a legal dispute, one side may be saying that the incident was due to cause A while the other side are saying that it was due to cause B. In fact, it is more likely to be due to both causes A and B, together with C and D which have not yet been discussed. The required judgement may therefore be a matter of assessing relative responsibility, rather than a clear division between the sides.

Forensic engineering: the investigation of failures. Thomas Telford, London, 2001.

Failures relate closely to accidents in terms of multiple causes and there is a study underway in the Institution of Mechanical Engineers which is surveying the causes of accidents. This is looking at large numbers of accidents and reports to try and identify the various generic categories of causes which can interact in various situations to result in an accident.

A summary of the conclusions from this study should help to avoid accidents in the future by making designers, operators and managers aware of the patterns. They should then be more aware of any critical situation where there is a build up of the factors that are ready to combine together, in a particular case, to give rise to an incident. Some typical generic categories are likely to be factors such as, design defects, installation problems, human factors such as tiredness or interpersonal relations, environmental factors, defects in materials or manufacture, system interactions, etc,etc.

FAILURES AND TOTAL LIFE COST

A difference between mechanical engineering failures and civil and structural engineering failures, is that the mechanical failures tend to arise from wear or fatigue, often after a long time in service. They therefore tend to be accepted as being to some extent inevitable, and part of a pattern of normal operation. A judgement is often made, actually or intuitively, on the basis of the total life cost of owning and operating a particular machine. If therefore a particular component or machine is available for half the price of another, it may be seen as acceptable, provided it lasts for well over half the time. Another important factor in component or machine selection is the failure pattern that it shows. The ideal pattern is one where failure occurs not only after a long time, but where the failure time bandwidth is narrow, so that reliable performance can be guaranteed for a long time in operation.

A feature of projects in civil and structural engineering is generally that the majority of the cost is incurred in the construction work and not so much in the subsequent operation of the facility. There is therefore a general tendency to try and keep the initial costs down to a minimum. With mechanical equipment however this can be very non-optimum, since total life cost is a much better criterion. If therefore mechanical equipment is being purchased as a part of a civil engineering project it is very important to recognise that different cost criteria need to be used.

CASE STUDIES

A good example of mechanical equipment as a part of a civil project is public service escalators. The engineering issue here is that the most critical components of an escalator, in terms of reliability, are the step chains, fitted at both sides of the steps, and used to drive them around the track. These chains are similar to very large bicycle chains and have several hundred pins and bushes between all their links. Because of their large number, it is essential that these components have a very consistent pattern of reliability. Experience over the last twenty years has shown that this can be achieved by the use of reinforced plastic bushes operating with stainless steel or chromium plated pins. These all wear very slowly and progressively and give a consistent service life of between forty and fifty years. In contrast the use of bicycle type steel on steel grease lubricated pins are prone to random failure after about five years. This arises because any tendency to local wear in any pin creates wear debris which clogs up the grease and inhibits further lubrication. Each of the hundreds of pins and bushes is therefore a failure waiting to escalate at any time, and creates a pattern of unreliability which requires a costly pattern of work to keep under any control. The problem is that use of the plastic bushes carries a cost penalty of about 10% and in a civil engineering purchase can easily be refused.

Similar problems have occurred on passenger lifts in applications such as railway stations. Here the lift operates, often almost continuously, for up to twenty hours per day, travelling between two levels only. The problem here is that the lift manufacturers experience is based largely on the installation of lifts in office blocks, hotels and blocks of flats. A characteristic of these applications is that the lift operates with a very random pattern of floor visiting, unlike the station application where the lift operates between two levels only. This means that the same part of the lift rope is subject to a large number of stress cycles assocated with each start from rest. It is therefore essential to use larger rope pulleys in these public service applications in order to get an acceptable rope life. This may seem obvious when the problem is explained, but it is easily forgotten when trying to achieve the lowest initial cost. It may also be an example of a multi-cause failure situation where the contributory causes are a non-standard operating cycle, a particular pattern of financial thinking, a good pattern of future replacement business for the suppliers and a lack of competent technical auditing by the purchasers.

In situations which are more mechanical engineering led, there tends to be more emphasis on total life costs. In power stations for example most failures arise from unexpected system problems such as the resonant vibrations discussed earlier, or the failure to recognise an environmental problem. For example in coal fired power stations the fuel is usually converted to a powder for ease of delivery and combustion in the boilers. Since the coal is stored outside the station, it is common to have the coal mills in an area exposed to outside temperatures. Problems have then occurred when large mills are cold started because their large casings remain cold for some time while the internal parts warm up more rapidly. This results in operating clearances being lost because a warm shaft in a cold housing gives a reduced clearance in its bearings This creates more heat and the process can become unstable.

Failures on most mechanical equipment can be avoided or controlled by condition monitoring. This is a process where regular measurements are made of indicators of safe operation, such as vibration levels or the level of wear debris in the machine lubricant. A practical problem in the management of these systems is that the operators become bored because the measurements show nothing of interest for a very long time. This can be overcome by automating the process but some degree of direct contact between the operators and their machines is still desirable. Some years ago a problem arose with the monitoring of the main rotor gearbox of a helicopter. These gearboxes cannot be designed to have an infinite life, within their weight limitations, and they are conventionally monitored with removable magnetic plugs to check for incipient failure debris from the gears and bearings. On this helicopter, the gearbox failed in flight and the investigation revealed that there had been a misunderstanding between the manufacturers and the operators. The operators wrote to the manufacturers with a question about the type of wear debris being generated. This letter went in error to the sales department who wrote back and said that the helicopter was very reliable and they should not worry. As a result the operators misinterpreted the allowable limit for the amount of wear debris of 50 sq mm, ie. a 7mm square as a 50 mm square, and the failure was a direct result of this error.

THE MAINTENANCE OF ROBUST EQUIPMENT
There is also an extremely important interface between the failure of engineering components and the management of companies which is worthy of a lot more attention. This arises because the objective of most companies is to maximise their profits at the end of the current financial year. One way of doing this is to study the expenditure budget of the company and see if it can be reduced. In many companies such as those operating public transport services,

the maintenance of their equipment infrastructure is one of their major items of annual expenditure. It is therefore very tempting to see if it can be cut back, and a common technique is to try a cut of 10% or 15% in any year and see if it has any effect. With the kind of robust equipment in use, a cut of this kind will generally have very little measurable effect in one year. The accountants on the board of management can therefore congratulate themselves, and consider that the results prove that, as they suspected, the engineers have been wasting money. They may then try further cuts, monitoring the results on an annual basis. The situation however is that with robust equipment it takes several years before the major effects of under-maintaining the equipment becomes obvious, by which time it is too late. Savings of a few million pounds a year for five years can require recovery costs running into hundreds of millions of pounds at the end of this period.

There is therefore a major need for analyses to be carried out of the sensitivity of various kinds of mechanical equipment to the effects of reduced maintenance. Data from the experience of the condition monitoring of equipment should be able to make a contribution to this study, but a fundamental investigation considering all the components of each machine and their possible interactions in various unreliability situations needs to be done.

Experience from risk analysis and the production of safety cases could certainly contribute to this study but the subject is wider than this. The information from such a study would provide a basis for obtaining economic availability of engineering equipment with an established and accepted method of decision making at the interface between accountancy, economics and engineering.

CONCLUSIONS

It is important to recognise that mechanical components within a civil engineering infrastructure require an adequate initial quality and sufficient maintenance to match their operating life to other parts of the system.

There is a need to review the reliability patterns of mechanical equipment, with an appreciation of their time response to inadequate maintenance, in order to avoid expensive failures.

Long life mechanical equipment has many interfaces with the civil engineering infrastructure, and there should be many ways in which this interface can be improved.

Structural failure and assessment of a department store

DR. M. HOLICKÝ
Klokner Institute of the Czech Technical University in Prague, Czech Republic

INTRODUCTION

The load bearing structure of a recently built department store in Prague consists of the flat (double ribbed) reinforced concrete slabs supported directly on columns located within span distances 12 × 12 m. Slabs above the first and second storey cantilever out by 3 m beyond the edge columns. After few years in service serious performance defects of cladding, interior partitions, and other secondary elements had been observed [1,2].

Incidentally at the same time another department store of the same user collapsed. This was perhaps partly the reason why all the performance defects of the new building had been carefully recorded and publicly reported, although the collapsed department store was a steel structure and its failure occurred due to well-recognised causes. Unfavourable engineering climate (psychological aspects) seems to have played a significant role in the subsequent assessment of structural damage. The observed defects were often exaggerated and interpreted as structural condition indicating insufficient safety against collapse, not just serviceability defects. Consequently, after less than 10 years in service, the second storey of the two-floor building was completely closed and the damaged non-bearing components were reconstructed.

A detailed analysis has shown that the serviceability failure of the second storey was primarily caused in design by lack of consideration of deflections from permanent load and shrinkage [3,4]. Presented theoretical model of public perception, which is a substantial extension of the previous study [4], takes into account uncertainty and vagueness in perceiving observed defects as well as effects of possible consequences. It appears that the developed theoretical model can well explain the wide differences between the public perception and the expert assessment of its structural condition.

DESCRIPTION OF THE STRUCTURE

The plan view dimensions of the building are 78 × 53 m. The load-bearing structure consists of reinforced concrete double ribbed slabs of the total thickness 0.45 m supported directly by columns of the cross section 0.5 × 0.5 m or 0.7 × 0.7 m located within span distances 12 × 12 m (see Figure1). The slabs are provided by in-depth heads, 1.65 × 1.65 m (at the edge columns) and 3.35 × 3.35 m (at the interior columns), where the coffer ceiling is replaced by solid slab. In the remaining part of the slab, the ribs of the cross section 0.18 × 0.38 m support a thin plate (of the thickness 0.07 m and interior spans 0.8 × 0.8 m). Equivalent thickness of a solid slab having the same rigidity would be 0.34 m only, which indicates that the stiffness of the slab is very low. Moreover, slabs above the first and second storey cantilever out beyond edge columns by 3 m.

Forensic engineering: the investigation of failures. Thomas Telford, London, 2001.

Design loads of the slab above the ground floor considered in the original analysis consist of a permanent part 7.0 kN/m^2 and a variable part 4.0 kN/m^2, corresponding values for the slab above the first floor are 7.2 kN/m^2 and 1.5 kN/m^2. However, the actual permanent load of this slab is due to actual roof and ceiling greater and could be as high as 10 kN/m^2. Nevertheless, the ultimate strength of the slab is sufficient, and no strengthening of the load bearing structure was needed.

Figure 1. Internal view of the structure

Non-bearing elements of the second storey, which were affected by deformations of bearing structures consist of façade cladding, interior partition walls and interior built-in components like glass walls and shelf stands. Cladding of the building consists of large glass windows, brick walls and window pillars, located within regular distances of 2.4 m. The window pillars, reinforced by steel ties anchored into floor and ceiling, were built in the bearing slabs without any movement joints. Similarly all interior masonry partition walls, reinforced by rolled steel elements of I- and U-section were constructed without any separation from roof slab. Expansion joints were not used in any interior built-in components. All of these non-bearing structures were evidently designed without desirable consideration of differential deflection of both floor and ceiling slabs due to their different stiffness and loading.

PERFORMANCE DEFECTS

After few years in service serious performance problems concerning cladding as well as interior non-bearing structures have been observed and analysed [3,4]. The most alarming were perhaps the defects appearing in the non-bearing structures: cracks of partition walls (see Figure 2 and 3).

Figure 2. Partition wall

Figure 3. Partition wall

Figure 4. Façade corner

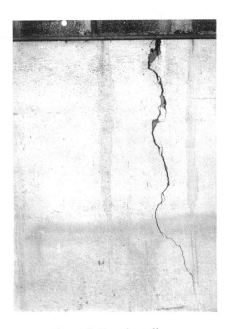

Figure 5. Façade wall

Tensile cracks of the window pillars and cladding elements were particularly noticeable near the exterior corners of the building (see Figure 4 and 5). Also, the deformed doorframes and a buckled shelf stand were visible and impressive. There were also sporadic shear cracks in the slab ribs close to the hidden column heads, however, they were mostly near construction joints. These cracks, which were of smaller importance for the overall damage assessment of the building, were the only detected defects of the load bearing structure.

ASSESSMENT

All of the defects of non-bearing structures in the first floor are caused by differential deflections of the floor slab (above the ground floor) and of the roof slab, primarily due to permanent load and shrinkage. As mentioned above the permanent load of the more flexible roof is by 3 kN/m^2 greater than that of the slightly less flexible floor slab. This could lead to considerable differences in the midspan deflection (shortening up to 30 mm) as well as cantilever deflection (extension up to 5 mm, at exterior corner up to 10 mm). Also shrinkage may lead to similar mutual differences in slab deflections; midspan shortening and cantilever extension may reach 5 mm (extension at exterior corners may be 10 mm).

Deformations due to temporary load and temperature are of smaller importance and may cause a maximum midspan extension 6 mm and shortening 1 mm, cantilever extension or shortening less than 1 mm. The observed deformation effects correlate well with theoretical results. Unfortunately, no calculation of structural deformations had been made in the original design. This was partly due to inadequate provisions of the contemporary standards.

Another construction fault mentioned above concerns the permanent load of the roof slab. Due to the actual self-weight of the ceiling and roof (sloped concrete layers) the permanent load of the slab is by 30 % greater than that assumed in design calculation. However, this discrepancy arises from the neglect of some load components in the design calculation. Also the actual permanent load of the slab above the ground floor may be slightly greater than that considered in design. Even though the deformation effects would be considerably lower if the actual load equalled the assumed load, many of the serviceability defects would have appeared anyway. This concerns, above all, the cracks in the cladding elements, particularly at the exterior corners of two-way cantilevered slabs. Most of the observed defects were primarily due to lack of consideration of deflection in the design and due to inaccurate determination of the design load. Construction errors, however, considerably added to deformation effects.

REPAIR

Although there were almost no defects apparent on the load bearing structure (except sporadic shear cracks in ribs), it was decided to close the first floor and to reconstruct damaged interior partitions and other non-bearing structural components as well as the cladding components. Interior partitions including steel elements have been separated from the ceiling and new movement joints have been covered by panel strip. New ties located near window pillars have mutually tied up the cantilevered slabs. These measures were proposed by the designer, even though an additional analysis shows that the expected deformations due to temporary load and temperature (as mentioned above) are limited, and in some cases could be admitted without any modification of the non-load-bearing structures.

PUBLIC PERCEPTION

Public perception (recently discussed in [5]) played an important role in the assessment and final decision concerning the building. The new department store became soon a building closely watched by a large number of users and local authorities. For more, at the same time another department store suffered from construction faults and this was partly the reason why all the performance deficiencies have been carefully recorded (similar experience is described in [5,6]). This unfavourable engineering climate seems to enhance the intensity of public perception. The observed defects were often exaggerated and regarded as indicators of insufficient structural safety. Widespread public perception of defects and discrepancies in expert assessments was reported in newspapers and finally resulted in a strong public demand for strengthening of the building.

THEORETICAL MODEL OF PUBLIC PERCEPTION

Evaluation of public as well as expert assessments has indicated that there is no distinct point in any commonly used performance indicator x (e.g. deflection, crack width) that would uniquely separate acceptable and unacceptable structures. Rather there seemed to be a transition region $<a, b>$ in which the structure gradually becomes unserviceable and the degree of caused damage $v(x)$ increases [4,7]. A conceivable model for $v(x)$ is indicated in Figure 6 as a trilinear function. Note that the values of $v(x)$ are within the conventional interval from 0 to 1. There is no damage below a certain lower limit value a, and full damage above the upper limit b.

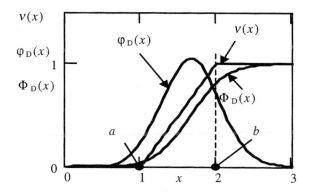

Figure 6. Perception model

Obviously, at any damage level $v(x)$ there may be a perception scatter expressed by the distribution function $\Phi_P(x,\mu_P,\sigma_P)$, for which lognormal distribution having the mean $\mu_P = x$ and standard deviation $\sigma_P = s \times a$, where s is a relative measure of scatter related to the lower bound a, is accepted here. Taking into account all levels $v(x)$, the cumulative damage $\Phi_D(x)$ is [4,7]

$$\Phi_D(x) = K\int_a^b v(\xi)\,\Phi_P(x,\xi,\sigma_P)\,d\xi \qquad (1)$$

where K is the normalising factor to normalise $\Phi_D(x)$ into the interval from 0 to 1 and ξ is the integration variable. The cumulative damage $\Phi_D(x)$ and a corresponding density function $\varphi_D(x)$, shown in Figure 6 for $a = 1$, $b = 2$ and $s = 0.3$, can be considered as generalised probabilistic models (involving economic aspects). The cumulative damage $\Phi_D(x)$ can be used to specify in a rational way the required serviceability constrains [2,3].

Further, considering an appropriate load effect E (e.g. deflection, crack width) having the mean μ_E and an assumed coefficient of variation w_E (=0.2) the expected perception level $\pi(\mu_E)$ can be defined as

$$\pi(\Delta,\sigma_P,\mu_E) = \int_{-\infty}^{+\infty} \varphi_E(x)\, \Phi_D(x)\, dx \tag{2}$$

Here $\varphi_E(x)$ is the probability density function of E. Gamma distribution having the mean μ_E is assumed. The mean μ_E equal to the lower limit of the transition region a, $\mu_E = a$, and the standard deviation $\sigma_E = 0.2\mu_E$ (the coefficient of variation 0.2) is assumed in the example shown in Figure 7, where the expected perception level $\pi(\Delta,\sigma_P,\mu_E)$ is indicated as a function of the ratio $\Delta = (b-a)/a$ for selected σ_P. Figure 8 shows $\pi(\Delta,\sigma_P,\mu_E)$ versus μ_E.

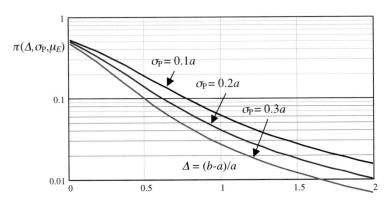

Figure 7. Perception level $\pi(\Delta,\sigma_P,\mu_E)$ for $\mu_E = a$.

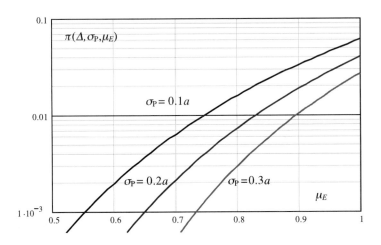

Figure 8. Perception level $\pi(\Delta,\sigma_P,\mu_E)$ for $\Delta = 1$.

It follows from Figure 7 and 8 that the expected damage (perception) level $\pi(\Delta,\sigma_P,\mu_E)$ is strongly dependent on both the ratio $\Delta = (b-a)/a$ and the mean μ_E, which may further depend on the sensitivity or experience of the observer. This finding explains the observed differences in public perception and discrepancies in expert assessments.

EFFECTS OF EXPECTED CONSEQUENCES

In addition to psychological aspects an assessment of a structure may be affected also by the expected economic (and other) consequences. Both these factors may be considered explicitly or implicitly only. On the one hand consideration of expected consequences may lead to a rational specification of structural constraints, on the other hand it may significantly affect the assessment of the structure. Moreover, the expert assessor may use uncertain data and information concerning the malfunctioning cost. The following analysis investigates the possible effects of expected consequences.

Assume that the optimum mean μ_E of the load effect E would correspond to the minimum of the total cost C_{tot}. Assuming a given coefficient of variation w_E, the optimum mean μ_E of the load effect E (the most suitable value assessed by an expert) may be obtained from the minimum of the total expected cost of a structure. Consider the total cost C_{tot} as a sum

$$C_{tot} = C_0 + C_m/\mu_E + \pi(\Delta,\sigma_P,\mu_E)\, C_f \qquad (3)$$

where C_0 denotes the initial cost independent of μ_E, and C_m/μ_E is an additional cost affected by modification of the mean μ_E, where C_m is a relevant marginal cost. Finally the multiple $\pi(\Delta,\sigma_P,\mu_E)\, C_f$ describes the expected cost due to structural failure. Obviously the minimum C_{tot} will occur when the partial cost

$$C_{part} = 1/\mu_E + \pi(\Delta,\sigma_P,\mu_E)\, C_f/C_m \qquad (4)$$

is the minimum, where $C_{part} = (C_{tot} - C_0)/\, C_f$. Figure 9 shows the partial cost C_{part} for the following parameters: $(b-a)/a=1$, $\sigma_P=0.3a$, $\sigma_E=0.2\mu_E$.

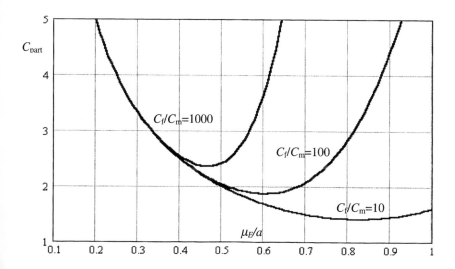

Figure 9. The partial cost C_{part} versus μ_E/a for $\Delta = (b-a)/a=1$, $\sigma_P=0.3a$, $\sigma_E=0.2\mu_E$

It follows from Figure 9 that the optimum mean μ_E may be expected from $0.5a$ to $0.8a$. However, for extremely high cost of failure C_f greater than 1000 C_m the optimum mean will be lower than $0.5a$. Obviously, depending on the assumed cost ratio C_f/C_m the optimum (the most suitable) mean μ_E of the load effect E may considerably vary. A further analysis is certainly needed in order to take into account various parameters and expected consequences, including life safety.

CONCLUSIONS
1. The serviceability failure of the department store was primarily caused by lack of consideration in design of deflection due to permanent load, shrinkage, variable loads including thermal actions.

2. The construction faults (additional permanent load and imperfect anchoring of ties) increased the unfavourable deformations of the structure.

3. The current engineering climate (psychological aspects) and the expected consequences seem to play an important role in the public perception of performance deficiencies, in the subsequent structural assessment and in the decisions made by the public authorities.

4. There is no distinct value that would uniquely distinguish acceptable and unacceptable structural conditions.

5. The disturbing variance in public perception and in expert assessment of the observed defects may be well explained using proposed theoretical model for public perception.

6. There seems to be an optimum value of the performance indicator that would lead to the minimum total cost and may be considered by an expert as most suitable.

7. The effects of vagueness in structural requirements and of the possible unfavourable consequences, including jeopardy of life, on the judgement of experts requires a more detailed investigation..

REFERENCES
[1] Holický M., Serviceability problems of department store - Case Study, Symposium on serviceability of buildings, NRC Canada, 1988, pp. 166-171.
[2] Holický M., Fuzzy concept of serviceability limit states. Symposium on serviceability of buildings, NRC Canada, 1988, pp. 19-25.
[3] Holický M., Fuzzy probabilistic optimisation of building performance, Automation in Construction, 8/1999pp 437-443.
[4] Holický M., Performance Deficiency of a Department store - Case Study. IABSE International Conference on Safety, Risk and Relibility - Trends in Engineering, Malta, March 2001, pp 321-326.
[5] Lewis R., The public perception of risk. RSA Journal, November 1995, pp. 52-63.
[6] Steward M.G. and Melchers R.E., Probabilistic risk assessment of engineering systems. Chapman & Hall, London, 1997.

ACKNOWLEDGEMENT
This research has been conducted at the Klokner Institute of the Czech Technical University in Prague, Czech Republic, as a part of the research project CEZ: J04/98/210000029 "Risk Engineering and Reliability of Technical Systems".

LIST OF SYMBOLS

a lower bound of the transition region
b upper bound of the transition region
s relative measure of the perception scatter σ_P/a
x performance indicator
w_E coefficient of variation of the load effect E
C_0 initial cost
C_f malfunction cost
C_m marginal cost
C_{part} partial cost
C_{tot} total cost
E load effect
K normalising factor
μ_E mean of the load effect E
μ_P mean of the perception level
σ_P standard deviation of the perception level
$v(x)$ damage level
$\varphi_D(x)$ probability density of the damage
$\varphi_E(x)$ probability density function of E
$\pi(\Delta,\sigma_P,\mu_E)$ perception level
ξ generic point of the performance indicator x
Δ transition region ratio $(b-a)/a$
$\Phi_P(x,\mu_P, \sigma_P)$ distribution function of the perception level
$\Phi_D(x)$ distribution function of the damage level

Forensic Evaluation of Concrete Industrial Slabs-On-Grade

S. M. TARR, Senior Engineer, and W. G. CORLEY, Senior Vice President,
Construction Technology Laboratories, Inc., Skokie, Illinois, USA

INTRODUCTION

Industrial slabs-on-grade are often subjected to demanding service conditions. To provide acceptable serviceability, the floor must be properly designed, constructed, and maintained for the intended use. As the working platform for industrial facilities, slabs with inadequacies can have severe impact on operations and can influence business profitability in areas including health and safety issues, equipment maintenance costs, rate of product shipping and receiving, and potential infestation.

CTL has conducted numerous forensic evaluations of industrial slabs-on-grade where performance has become a concern. Performance issues such as random cracking, joint spalling, excessive slab curling, surface wear, surface delaminations, and differential slab settlement can impact the industrial processes for which the slab was designed. Slab failures cause the user of the industrial facility to alter standard operating procedures to accommodate the slab distress. Examples of such procedural modifications include decreased lift truck speed, decreased lift truck load, routing lift truck traffic around slab defects (aisle closures), and increased equipment maintenance and related down time.

When floor problems arise, they often involve a tenant, owner, architect, and contractor. Often, the architect engages a structural engineer to design the slab-on-grade and the general contractor will often subcontract a concrete specialty firm to place the floor. When the problems cannot be resolved within these parties, the assistance of experienced consultants is often required to determine the cause of the distress and facilitate an acceptable resolution.

BACKGROUND REVIEW

The first step in evaluating the cause of a slab-on-grade failure is to perform a thorough review of pertinent background information. Pertinent project-related documentation includes subsurface soils explorations of the site prior to construction, design calculations, construction weather reports, construction specifications and plans, concrete mix design, concrete quality assurance data, and daily field inspection logs.

The subsurface soils exploration report provides useful information pertaining to subgrade conditions if the type of slab failure involves differential settlement or expansive materials. Most quality geotechnical reports include engineering properties of the insitu soils encountered and recommendations for the modulus of subgrade reaction which should be used in the slab structural design.

Forensic engineering: the investigation of failures. Thomas Telford, London, 2001.

Design calculations should be reviewed to determine the level of serviceability provided by the design. Specific design input parameters such as lift truck loads, distributed load layout, and rack configuration should be verified for accuracy.

Construction specifications and plans should be reviewed.   Details provided by the specifications such as slab thickness, joint spacing, reinforcement, and joint depth can be easily verified during the evaluation. Other parameters such as concrete mix design, finishing technique, and curing procedures can also be assessed through additional testing.

Other records such as weather data during construction, construction inspection reports, and quality control/assurance data should also be reviewed. These documents may provide useful information depending on the type of slab failure exhibited.

VISUAL ASSESSMENT

A visual inspection of the industrial facility floor is made to collect data about the specific failure. Some information collected can be used to make definitive conclusions regarding the cause of the failure. For example, if joints are left open instead of being filled with a semi-rigid joint filler, they can be expected to spall under hard, solid-rubber lift truck wheels. Joint spalling is the chipping of slab edges due to high shear stresses when trafficked by solid wheels. A flexible joint filler (caulking) does not provide adequate support to joint walls to prevent spalling and a rigid epoxy does not allow for minor slab expansion and contraction such as caused by temperature fluctuations. If the joints appear to be properly filled with a semi-rigid filler and are still spalling in traffic areas, it is possible that a backer rod was used. A backer rod is used in exterior pavement joints subjected to pneumatic tires which are filled with a flexible sealant. The backer rod is inserted in exterior joints to provide the proper aspect ratio (depth to width) for flexible sealants. However, if a backer rod is used on interior joints subjected to hard-wheeled lift trucks, shear stresses may still cause spalling to occur at a width equal to the depth of the backer rod. For this reason, a semi-rigid joint filler should be installed to the full depth of sawcuts installed at control joints and construction joints. And even if the joints were properly filled full-depth with an appropriate semi-rigid epoxy, additional slab shrinkage can cause the filler material to split (cohesion epoxy failure) and/or separate from the joint edge (adhesion epoxy failure). When this occurs, joint wall support is no longer provided and spalling can occur. Adhesive and cohesive epoxy failures should be repaired by re-installing filler material.

The following additional observations are generally made to assess the concrete slab as-placed and to provide direction for additional testing during the evaluation:

- Construction and control (contraction) joint spacing and condition is documented. According to agencies such as the American Concrete Institute (ACI) and the Portland Cement Association (PCA), joints should be installed at a frequency (in feet) of 2 to 3 times the slab thickness (in inches). Therefore, for a 6-in.-thick slab, the proper joint spacing should be 12 to 18 ft. In our experience, to minimize the potential for random cracking, the maximum joint spacing should be limited to 30 times the thickness (15 ft for a 6-in.-thick slab). The joint layout should be designed based on the column layout. Joints should occur along all column lines and the number of joint between columns should be installed based on the above rule of thumb. If the column spacing is different for the two directions, an attempt should be made to maintain slab panels as square as possible (maximum length-to-width ratio of 1.25).

- Crack patterns noted provide clues to their cause. While cracks can be caused by the loading applied to the slabs (especially for insufficient thickness), they can also be the result of design and construction flaws. To attribute causation to specific cracking, additional data is commonly needed. However, repetitive occurrences of the following types of cracking may be related to design and construction inadequacies.

  - Mid-panel cracks can be a sign that the joint spacing is excessive or installed control joints are not working properly (not cracked below sawcut). Even if sawed joints are properly spaced, the joint may not be deep enough to create the weakened plane necessary to "control" the location of cracking. Joints installed to the proper spacing and depth before tensile stresses are developed which exceed the tensile strength attained by the concrete minimize random, out-of-joint cracking. According to ACI and PCA, the depth of the sawcut should be ¼ the slab thickness. In addition, if control joints include dowel bars, attention should be given to their alignment. Dowel bars chaired on baskets can easily become misaligned during construction. Misaligned dowels do not allow intended joint movement (opening during initial slab contraction). Consequently, random cracking occurs. Also, continuing temperature reinforcement (welded wire fabric or deformed bars) through sawed control joints may prohibit joint opening (cracking below sawcut) and result in random uncontrolled cracking.

  - In plain concrete slabs, cracks occurring adjacent to a sawed joint can be indicative of late joint sawcutting. Studies have shown that the optimum time for sawcutting is when the peak heat of hydration is reached (generally 8 to 12 hours after placement). After the maximum hydration temperature is reached, the concrete temperature drops causing thermal contraction in addition to early age drying shrinkage. This is especially true when slabs are constructed without the protection of the building envelope (prior to constructing the walls and roof) in areas with large daily temperature fluctuations.

  - Corner breaks can be caused by loading of unsupported corners due to insufficient thickness or curling and warping of slab panels.

  - Crazing cracking (map cracking) is a network of fine cracks in a chicken-wire pattern that do not penetrate far below the surface. Crazing is indicative of early surface drying which can occur when the slab is improperly cured.

- Significant warping and intermittent surface unevenness should be noted. Even when flatness criteria are not specified, ACI guidelines for proper slab finishing tolerances should apply. If the slab has curled, excessive joint movement under loading is likely. Any vertical movement should be documented. Movement may be detected by standing on joints when a lift truck passes or by a thumping sound at joint locations when the lift trucks traversed the slab.

- All floor surface defects are noted. These may include dusting, blistering, popouts, discoloration, delaminations, excessive wear areas, and insufficient surface slip resistance. Various observations will prompt further testing. For example, if the floor is slippery, surface friction testing is performed in accordance with ASTM F-609 *Standard Test Method for Static Slip Resistance of Footwear Sole, Heel, or Related Materials by Horizontal Pull Slipmeter (HPS)* to verify compliance with Occupational Safety and Health Administration (OSHA) and Americans with Disabilities Act (ADA) requirements. Delaminations or excessive wear indicate concrete mix or finishing problems and require petrographic examinations. The relative surface hardness of the slab is qualitatively assessed in the field. Petrographic examination can quantify the depth of a relatively soft surface prone to dusting.

- The presence of isolation joints along walls and at slab penetrations (column footings, bollards, etc.) should be noted. Slab-on-grade movement should be isolated from rigid structural components and other features which penetrate the slab in order to minimize the occurrence of re-entrant cracking.

NON-DESTRUCTIVE TESTING

Ground Penetrating Radar
Nondestructive ground penetrating radar (GPR) testing can be used to evaluate parameters such as slab thickness, dowel bar misalignment, reinforcement location, and sub-slab voids. Commercially available GPR operates on the principles of electromagnetic wave reflection. A small 100 MHz transducer transmits electromagnetic pulses into the concrete. The pulses are partially reflected back to the transducer by interfaces between materials of different dielectric constant (concrete and air, concrete and steel, concrete and subgrade material, etc.). The remaining portion of the pulse is refracted into the next medium, where more of the pulse can be reflected. Reflected pulses are received by the transducer, electronically processed, and digitally recorded. The location of an object (plan and elevation) is determined by the amount of time between pulse transmission and reception.

When slab thickness is a concern, GPR can be used to rapidly determine the thickness of accessible portions of the slab. As cracking is most likely to occur in areas of minimum thickness, the GPR data can be summarized to provide the minimum thickness of each slab panel. This information is useful when developing repair quantities.

Proper installation of steel reinforcement can also be verified using GPR. Depth and variation of temperature steel can be assessed and reported. In slabs-on-grade, temperature steel is often specified to hold randomly occurring cracks tight. Temperature reinforcement does not prevent cracking. However, if cracking occurs away from the intended locations (joints), the steel holds the crack tight providing aggregate interlock load transfer which decreases required crack maintenance. As outlined in ACI 302, "temperature and shrinkage cracks in unreinforced slabs originate at the surface of the slab and are wider at the surface, narrowing with depth. For maximum effectiveness, temperature and shrinkage reinforcement in slabs-on-grade should be positioned in the upper third of the slab thickness." The Wire Reinforcement Institute recommends that welded wire reinforcement be placed 2 in. below the slab surface. Steel mesh located at or near the bottom of the slab, as regularly observed, is ineffective in holding cracks tight. While GPR has been used to verify the presence and location of specified temperature steel, most properly designed slabs on grade can be constructed without welded wire fabric as long as joints are properly spaced and sawed in a timely manner to a depth of ¼ the slab thickness.

GPR can also examine the presence and alignment of dowels bars. Control and construction joints can be surveyed for horizontal misalignment, vertical misalignment, offset (dowel bars positioned to one side of the joint), and for missing dowel bars. Dowel bar misalignment surveys are performed using two antennas spaced 12 in. apart (each antenna scanning parallel paths 6 in. from joint). Dowel bar misalignment greater than ¼ in./ft in either the horizontal or vertical direction is typically considered excessive.

Impulse-Response
When slab failures involve curled/warped slab edges or delaminated surfaces, it is often helpful to use Impulse-Response Testing to quantify the extent of the failure and the magnitude of the required repair. This method uses a low strain impact to send a stress wave through the tested element. The impactor is generally a 1-kg sledgehammer with a built-in load cell in the hammer head. The response to the input stress is normally measured by a velocity transducer (geophone). Both the hammer and the geophone are linked to a portable field computer for high speed data acquisition and storage. The records (hammer force and geophone velocity) are processed (velocity spectrum divided by force spectrum) into a transfer function referred to as the mobility of the test element. The test graph of mobility plotted against frequency over the 0 to 1 kHz range contains information on the condition and the integrity of the slab. Portions of the plot define the dynamic stiffness and the dampening ability of the concrete slab. These parameters are used to determine the extent of the following slab-on-grade problems:

- Sub-slab voids (curled slab edges)
- Surface delaminations
- Low density concrete (honeycombing)
- Cracking
- Stress transfer across joints

Benkelman Beam
When slab movement at joints is a concern, deflection testing can be performed at locations selected to include a range of joint conditions (filled/unfilled, cracked/uncracked) and widths. Loaded lift trucks in service at the facility can be used as the load for the deflection testing. Deflection measurements are collected using precision digital dial gages attached to the end of a steel beam (Modified Benkelman Beam). Two dial gages are attached to the beam to facilitate the measurement of deflection load transfer across joints. The telescoping beam is supported outside the perimeter of the load deflection basin to ensure that the load does not affect the deflection measurements. The telescoping portion of the beam is extended to position the dial gages near the test location.

Often, excessive slab deflection is the result of deformed slab edges. The industry currently refers to the deformation of slab edges, regardless of the cause, as slab "curling." To clarify the issue, as the proper repair may be dependent upon the specific cause, slab "curling" is the deformation of the slab surface due to a difference in temperature between the slab surface and bottom. Like most materials, concrete expands and contracts with a change in temperature. If the slab surface is cooler than the slab bottom, the surface contracts causing the slab edges to curl upward. In contrast, slab "warping" is the deformation of the slab surface due to a difference in moisture between the slab surface and bottom. In much the same manner as a sponge, if the slab surface is allowed to dry and the bottom is kept saturated, the edges will tend to warp upward. In most interior slabs-on-grade, the edges of the concrete slab panels have, technically, "warped" upward due to a moisture difference between the slab top and bottom. When slab deformation is observed, the degree of warping should be quantified at deflection measurement locations. This can be done using a leveled straightedge extending between the slab center and the edge/corner or by the use of precise laser levels.

In addition to the warping and deflection measurements, joint widths should also be recorded. In general, sawing equipment installs a joint which is 1/8-in.-wide. As the slab contracts due to drying shrinkage, the joints open. The magnitude of the joint opening is dependent upon the joint spacing and shrinkage potential of the concrete mix. The joint width can be measured using a crack comparator at the time of the deflection testing. In our experience, when a joint opens in excess of 0.035 in., there is a loss of load transfer through aggregate interlock. If dowels are not installed in joints, there is minimal transfer of the load from slab to slab as the lift truck traverses a joint. This does not present a problem if the slab is designed accordingly and a sufficient thickness is provided.

CONCRETE CORE EXAMINATION
Cores are sampled from the slab-on-grade for a number of reasons including calibration of non-destructive testing data, petrographic examination of concrete quality, and verification of in-place concrete strength.

While non-destructive testing techniques can be used to rapidly assess the thickness of large areas of the slab-on-grade, these techniques need to be calibrated with core samples removed from the slab. This is due to the variability in the wave transmission rate through different concrete mixtures. The size, quality, and abundance of aggregate, as well as the quality of the cement paste, will impact the wave velocity. Core lengths are measured in accordance with American Society for Testing and Materials (ASTM) C-174 *Standard Test Method for Measuring length of Drilled Concrete Cores.*

If the quality of the concrete is a concern, core samples can be microscopically examined. The examination is performed in accordance with ASTM C-856 *Standard Practice for Petrographic Examination of Hardened Concrete.* Typical findings of the Petrographic examination are:

- General condition of material
- Causes of inferior quality, distress, or deterioration
- Probable future performance
- Compliance with project specifications
- Description of concrete providing:
  - Degree of cement hydration
  - Estimation of water-cement ratio
  - Extent of paste carbonation
  - Presence of flyash and estimation of amount
  - Extent of corrosion of reinforcing steel
  - Identification of evidence of harmful alkali-aggregate reaction, sulfate attack, or other chemical attack
  - Identification of potentially reactive aggregates
  - Evidence of improper finishing
  - Estimation of air content
  - Evidence of early freezing

Compressive strength of the in-place concrete can be measured using core samples. Testing is performed in accordance with ASTM C-42 *Standard Test Method for Obtaining and Testing Drilled Cores and Sawed Beams of Concrete.* While cylinders are often sampled during slab construction, these are generally cured in a laboratory moisture chamber. This environment provides excellent curing for cylinder samples, but this curing may not be applied to the slab. Therefore, while cylinder testing may indicate sufficient concrete strength, as-placed slab strength may not be equivalent. As provided by ACI 318, 3 cores should be tested for each 5,000 ft^2 when assessing the in-place strength of a single placement.

Cores can also be removed to visually examine other parameters such as whether control joints are working (cracked below sawcut), whether joint filler was installed full depth of the sawcut, and the effectiveness of repairs (epoxy penetration, bond strength, etc.).

DESIGNED VS. CONSTRUCTED LOAD-CARRYING CAPACITY
Invariably, the parameters of the slab-on-grade as-constructed will not match exactly with those designed. Aside from obvious surface defects which impact facility operations, the most critical parameters impacting the slabs structural functionality and overall service life are thickness, concrete strength, and modulus of subgrade reaction. In an effort to quantify the impact of deviations from the specified parameters, slab stress analyses for both the as-designed and as-built floor conditions are conducted. The floor is structurally evaluated to establish the relationship between the as-designed and as-built load carrying capabilities.

Design parameters are available from the construction specifications and drawings. As-constructed parameters are collected from construction records, quality assurance testing, and data collected during the field evaluation. The most common strength specification for slabs-on-grade is compressive strength. Published relationships are available to compute associated strength characteristics from the compressive strength. For example, the concrete modulus of elasticity can be computed using Equation 1 as provided by ACI 318.

$$E_c = 57,000 * (f'_c)^{1/2} \quad\quad\quad\quad ..... \text{Eq. 1}$$

Where E_c = concrete modulus of elasticity, psi

f'_c = concrete compressive strength, psi

The concrete modulus of rupture can also be estimated from the compressive strength. As-designed and as-built concrete flexural strengths are calculated using Equation 2, as published by the Portland Cement Association (PCA).

$$MR_c = 9 * (f'_c)^{1/2} \quad\quad\quad\quad ..... \text{Eq. 2}$$

Where MR_c = concrete modulus of rupture, psi

f'_c = concrete compressive strength, psi

Based on information provided by the geotechnical investigation, the modulus of subgrade reaction can be estimated. If an effective k is not directly provided by the geotechnical report, the design bearing value or California Bearing Ratio can be used to estimate the k using the Portland Cement Association publication PCA Soil Primer which includes the relationship of the most commonly measured soil characteristics.

The finite element computer program ILLISLAB, developed in 1977 for the Federal Aviation Administration (FAA) for structural analyses of concrete pavement systems, can compute load induced flexural stresses. The program is based on medium-thick plate on a Winkler (spring) foundation bending theory. It is capable of computing stresses and deflections for panels with doweled, keyed, or aggregate interlock load transfer at the joints. Multiple loads can be modeled at any location on a jointed or cracked system of slabs-on-grade.

Lift truck and rack loads are analyzed to determine the critical loading condition affecting slab performance (i.e., lift truck loads positioned at joints or cracks). The PCA thickness design method recommends designing slabs for loads that will not result in a stress ratio (flexural stress to flexural strength) that exceeds 0.50. The factor of safety of 2.0 (inverse of stress ratio) is recommended where heavy load traffic is frequent and channelized (aisleways and staging areas). A stress ratio of 0.50 allows an unlimited number of load repetitions (endurance level of concrete resulting in no fatigue load cracking).

For stress ratios exceeding 0.45, PCA developed fatigue equations based on Miner's Hypothesis of material fatigue to calculate the allowable number of load repetitions before fatigue failure (cracking). The equations have the following form:

For SR > 0.55
$$\mathrm{Log}_{10}(N) = (0.97187 - SR) / 0.0828 \qquad \qquad \ldots.. \text{Eq. 3}$$

For $0.45 \leq SR \leq 0.55$
$$N = (4.2577 / (SR - 0.43248))^{3.268} \qquad \qquad \ldots.. \text{Eq. 4}$$

For SR < 0.45
$$N = \text{unlimited} \qquad \qquad \ldots.. \text{Eq. 5}$$

Where SR = stress to strength ratio
N = number of allowable load repetitions

Using these fatigue equations, an analysis of both the as-designed and as-built floor load carrying capability is performed for the in-service loading conditions. Table 1 summarizes a typical non-reinforced concrete slab load analysis for a common lift truck. For this example, the design slab parameters were a concrete thickness of 8 in., a compressive strength of 4,000 psi, and an effective subgrade modulus of 150 pci. As shown in the table, the slab parameter having the greatest impact on load carrying capability is concrete thickness. The number of allowable load repetitions is significantly affected by slab underthickness. In this example, the allowable number of load repetitions is significantly reduced by over with a thickness reduction of only ½ in. This supports the importance of controlling thickness. ACI 117 recommends a thickness tolerance of –¼ in. and +3/8 in. The positive (overthickness) tolerance is recommended to minimize keying into the base often caused by tire ruts from concrete delivery trucks. Abrupt thickness inconsistencies increase the slab/base drag coefficient and may increase the occurrence of drying shrinkage cracking.

Table 1 - Example Analysis* of Designed vs. Constructed Slab Load Carrying Capability.

Analysis Condition	Concrete Strength Compressive, psi	Flexural, psi	Subgrade Modulus, pci	Slab Thickness, in.	Maximum Stress, psi	Stress Ratio	Allowable Repetitions
As-Designed	4,000	569	150	8.00	257	0.45	47,803,754
As-Built Underthick	4,000	569	150	7.50	280	0.49	1,155,305
				7.00	315	0.55	113,241
				6.50	355	0.62	16,043
				6.00	400	0.70	1,780
				5.50	460	0.81	95
As-Built Low Strength	3,500	532	150	8.00	252	0.47	3,947,159
	3,000	493	150	8.00	250	0.51	547,889
	2,500	450	150	8.00	247	0.55	128,381
As-Built Poor Base	4,000	569	100	5.25	267	0.47	5,636,762
	4,000	569	75	5.00	276	0.48	1,743,009
	4,000	569	50	5.00	285	0.50	736,285

* 12,000 lb lift truck modeled with 80 percent of total load on front axle.

After slab thickness, the parameter having the highest impact is concrete strength. However, a considerable number of load repetitions are still allowable with a significant strength reduction of 25 percent. In our experience, this magnitude of strength decrease is rare. And, as strength is generally monitored during construction, strength problems are usually discovered early and corrected during construction. More commonly, the as-constructed strength exceeds the design strength. This additional concrete strength must be incorporated into the analysis. However, slab underthickness dominates the analysis. In the above example, a significant concrete compressive strength increase of nearly 2,500 psi (6,500 psi) is required to offset a thickness reduction of just 1 in. The modulus of subgrade reaction has an even lower impact on load carrying capability as shown in the table. For this reason, it is seldom justified to expend the amount of effort required to test the as-constructed subgrade modulus. However, if significant slab settlement is observed, a sub-slab investigation is often necessary to attribute specific cause for the failures.

This analytical technique can also be used to evaluate a reduction in service life for curled/warped slabs. For this, Benkelman Beam deflection measurements are required to calibrate the ILLISLAB model as curled slabs tend to settle into the base which increases slab support. To calculate an accurate number of allowable load repetitions for curled slabs, the stresses in the slab caused by curling and loading are computed separately and added. This approach is consistent with plate theory and superimposing stresses. When slab edges curl or warp upward, stresses occur due to the slab weight alone. These "unloaded" stresses can be added to stresses computed in loaded slabs where the support strength was adjusted to result in deflections consistent with those measured during the site visit. This analysis results in accurate "as-built settled" stresses and associated allowable load repetitions. It should be noted that curled slabs create cantilevered edges that, when loaded produce significant stresses. These stresses and related impact on allowable load repetitions can be similar to the thickness reduction of 2.5 in. shown in Table 1. Therefore, if slab curling isn't corrected, the length of the floor's serviceability life can be substantially reduced.

SUMMARY AND CONCLUSION

Based on the combined findings of the evaluation, conclusions relative to slab serviceability are reported and recommendations for restoring the floor to the as-designed condition are developed. Required repair protocol to restore serviceability are dependent on the failures observed but can range from complete removal and replacement to structural epoxy injection of cracks to slab subsealing and surface grinding of curled/warped slab edges. Restoring serviceability may also include partial-depth repairs of badly spalled joints and application of a penetrating sealer/hardener to moderately soft surfaces. However, in all cases, proper repair procedures can only be developed once the cause of the distress has been determined.

The cost to perform long-term repairs to industrial slabs-on-grade can be significantly higher than the cost of original construction. This is especially true when considering the cost of the impact of repairs on the operation of the facility. The temporary relocation of product or manufacturing equipment and personnel may be required and these interruptions can have a large impact on business profitability. However, as the floor is the working platform for the facility, repairs are usually necessary. For this reason, and because the cause of the slab failures are not generally obvious, a complete evaluation, such as that described in this paper, is often required to settle a dispute between all parties involved.

REFERENCES

1. ACI Manual of Concrete Practice, American Concrete Institute, Farmington Hills, MI, USA, 1998.

2. "Concrete Floors on Ground," Engineering Bulletin No. EB075.02D, Portland Cement Association, Skokie, IL, USA, 1990.

3. Packard, Robert G., "Slab Thickness Design for Industrial Concrete Floors on Grade," Information Sheet No. IS195.01D, Portland Cement Association, Skokie, IL, USA, 1976.

4. "PCA Soil Primer," Engineering Bulletin No. EB007.05S, Portland Cement Association, Skokie, IL, USA, 1992.

5. "Thickness Design for Concrete Highway and Street Pavements," Engineering Bulletin No. EB-109.01P, Portland Cement Association, Skokie, IL, 1984.

6. Miner, M.A., "Cumulative Damage in Fatigue," American Society of Mechanical Engineers Transactions, Vol. 67, 1945.

7. Annual Book of ASTM Standards, American Society for Testing and Materials, Conshohocken, PA, USA, 2001.

8. Khazanovich, L. and Ionnides, A.M., "Finite Element Analysis of Slabs-On-Grade Using Improved Subgrade Soil Models," Proceedings, ASCE Specialty Conference 'Airport Pavement Innovations – Theory to Practice,' Waterways Experiment Station, Vicksburg, MS, USA, 1993, pp. 16-30.

Failure prevention in the construction project

ILIAS ORTEGA, DR.
Ortega & Kanoussi Technologies, Mexico City, Mexico

ABSTRACT

This paper discusses several methods for preventing construction failures: statistical investigations, case study research, transfer of failure prevention methods from risk-conscious industries to construction, peer reviews, monitoring, and redundant design.

INTRODUCTION

Construction failures have important social and economic effects: for example, in the U.S., approximately 450 persons die annually due to construction failures, while about 1,500 persons lose their lives on account of construction accidents (Eldukair and Ayyub, 1991). The direct and indirect costs of construction failures in the U.S. are estimated at about 15 billion USD. This means that about 5% of the worth of the U.S. construction projects is lost due to construction failures. A similar percentage has been estimated for the Swiss construction sector (Schneider, 1997). These figures underscore the need for measures to prevent construction failures. Furthermore, considering that construction projects have returns on investment of only a few percentage points, the profitability of construction projects can be significantly increased by preventing construction failures.

According to a study carried out at the Swiss Federal Institute of Technology in Zurich (Matousek and Schneider, 1976), approximately 75% of construction failure cases can be attributed to human error; while the remaining 25% is due to consciously accepted risks. An even larger part of the damage sum due to construction failures, about 90%, is due to human error; while the remaining 10% are attributed to consciously accepted risks. Therefore, the prevention of construction failures is not so much a technical issue, but rather a management issue.

This paper discusses several practical methods for preventing construction failures. The methods discussed belong to the following categories:

1. *Learning from failure.* This *data-oriented* approach is based on the study of construction failures in order to prevent similar failures. Statistical investigations and case study research are part of this approach
2. *Learning from success.* This *method-oriented* approach intends to transfer effective failure prevention methods applied in industrial sectors with high safety standards to the construction sector
3. *Redundant design.* This method verifies the plausibility and validity of design results through the *redundant* use of design approaches such as manual and computer calculations, experiments, and monitoring

Forensic engineering: the investigation of failures. Thomas Telford, London, 2001.

4. *Peer reviews.* Peer reviews are independent inspections of designs with the intention of detecting errors or potential problems in the reviewed designs
5. *Monitoring.* The purpose of monitoring is to determine if built structures behave as designed

LEARNING FROM FAILURE

The study of construction failures is valuable, since each failure provides information that can be used to prevent similar failures. Furthermore, the investigation of construction failures points to gaps in the theory and practice of construction and therefore triggers innovations. The investigation of construction failures can be divided in two broad categories:

1. *Statistical investigations* done on large, representative samples of construction failures
2. *Case study research*, i.e. in-depth investigations of individual construction failures

Statistical Investigations

Statistical investigations can lead to the discovery of patterns in construction failures such as the failure types occurring more frequently, the most common failure causes, etc. Thus, statistical investigations help determine the *structure* of construction failures (Matousek and Schneider, 1976). Besides describing the structure of construction failures, statistical investigations lead to measures of failure prevention such as indicating at which points in the construction process inspections are most effective. The main shortcomings of statistical investigations are large data samples, biased data, and limited access to data.

The *Ishikawa diagram* is an example of a statistical tool widely used to analyze the causes of defects in mass production (Juran and Gryna, 1993). However, Ishikawa diagrams can also be effectively used to analyze the causes of construction failures. *Fig. 1* shows an Ishikawa diagram based on data taken from Matousek and Schneider (1976). It is based on 295 failures traced to errors incurred during the design phase. The figure shows that about 50% of the damage sum is attributable to errors involving static calculations; therefore, the diagram suggests that major improvements in design can be achieved by preventing errors occurring in static calculations. An additional Ishikawa diagram carried out on the cases concerning static calculations can help discover in more detail the types of errors arising while doing calculations. Potential types of calculation errors are computer misuse, wrongful assumptions of load cases, etc.

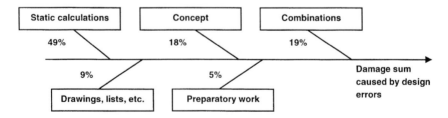

Fig. 1. Ishikawa diagram of design errors (data extracted from Matousek and Schneider, 1976)

Case Study Research

Ideally, case study research (Yin, 1994 and 1993) is done after statistical investigations have determined the industry-wide structure of construction failures. Then, the sample used for the statistical investigations is used to select construction failure cases with representative causes or effects for the more detailed case study research. The lessons learned from the individual failure cases can be analyzed, generalized, and used to prevent further failures. The main shortcomings of case study research are non-representative samples, biased data, high costs of research, and limited access to data.

An example of a construction failure offering many valuable lessons is the collapse of the ceiling of the Uster (Switzerland) indoor swimming pool on May 9 1985 (*Fig. 2*). Twelve persons, most of them children, lost their lives due to this tragic failure. The Uster failure provided several universally valid lessons such as recommending structural redundancy to prevent progressive failure and monitoring strategies to observe the behavior of structures. In addition, the failure had radical effects on Swiss construction practice and triggered many activities to prevent a similar tragedy. Among them, technical information was quickly disseminated, building codes were updated, and new building codes were developed.

Fig. 2. Collapsed ceiling of the Uster indoor swimming pool (Faller and Richner, 2000)

LEARNING FROM SUCCESS

The transfer of effective failure prevention methods used in risk-conscious industrial sectors to the construction sector can help prevent construction failures. An example of such a method is *incident (or near miss) reporting* (Schaaf, van der, et al. 1991), a method widely used in aviation. Additional examples of areas from which failure prevention methods can be transferred to construction are the following:

1. *Air traffic control.* Despite the large number of airplanes landing and departing each day from major airports, the accident rate at airports is extremely low. To a large extent, this is due to the communication procedures used by air traffic controllers
2. *Aircraft carriers.* The safe landing of airplanes on aircraft carriers is owed mainly to the reliable work and organization of the carrier crew. Such organizations performing high on the reliability scale are called *high-reliability organizations* (HRO) and have been the subject of extensive study (Pool, 1997 and Roberts, 1993). Features of

HROs, such as continuous monitoring, emergency planning, forgiveness if errors are committed, etc., can be directly transferred to operations on the construction site

Further areas from which failure prevention methods can be transferred to construction are navigation, space travel, management of chemical plants, management of emergencies in hospitals, etc.

PEER REVIEWS

Research (Matousek and Schneider, 1976) has shown that about 60% of the damage sum due to construction failures attributed to human error could have been prevented by additional timely inspections, while about 25% could have been prevented without any additional inspections. Therefore, in order to avoid incurring unnecessary risks and to increase the safety of innovative or unconventional designs, it is advisable to carry out thorough independent *design peer reviews* (Bell et al., 1989 and Zallen, 1990), even if they are not mandatory or even when the client refuses to carry their costs.

In Germany, design peer reviews are carried out by *Prüfingenieure* (reviewing engineers), who are government-licensed experts. Prüfingenieure determine if a particular design fulfills current building regulations. According to a study by Baur (1981), the costs of peer reviewing a typical building amount to about 1% of the structural costs. According to the same study, the costs of the damage prevented by peer reviews amount to about 3% of the structural costs. Errors in detailing drawings make up the error category most frequently detected during peer reviews. This error category is followed by errors such as incomplete loading assumptions, insufficient stiffness, and inadequate structural systems. The experience gained in Germany with peer reviews has shown so far the following benefits (Baur, 1981):

1. The requirement to thoroughly design the entire structure, including detailing, helps speed up the construction process and avoid expensive improvisations
2. Peer reviews help maintain a minimum quality level and prevent unfair competition
3. Peer reviews help benefit from the strength of construction materials. In Germany, safety factors are determined under the assumption that peer reviews are carried out. If peer reviews are eliminated, safety factors may have to be increased
4. Peer reviews help detect errors occurring at the interfaces of the design process
5. Peer reviews help promote compliance with building codes

MONITORING

Tragic failures such as the Uster collapse could have been prevented if the requirement to monitor critical structural elements had been identified during design. In addition, innovative or unconventional structures that have not yet been empirically validated, are particularly prone to unexpected failure mechanisms. Therefore, in order to determine if these structures behave as predicted, during construction and operation, a monitoring strategy should be developed during the design stage, before construction begins.

Fig. 3 shows a conceptual framework for monitoring strategies for construction. The construction process is simplified by considering three phases: *design, build* and *operate*. The feedback from monitoring during construction can be used to guide the construction process and carry out the necessary changes during construction (*feedback loop 1*). Feedback from monitoring during operation can help determine if the design assumptions were correct and can thus lead to changes in operation or even demand measures such as reinforcement (*feedback loop 2*). Finally, the monitoring experiences from construction and operation may result in design improvements for similar, future structures (*feedback loop 3*).

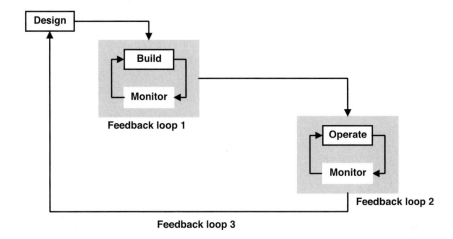

Fig. 3. Framework for monitoring strategies during construction

The complexity of monitoring strategies varies widely. The simplest monitoring strategy is a visual inspection, during which the overall condition of a structure is assessed with nothing more than the human eye. Nowadays, modern technology allows to monitor a wide spectrum of parameters such as displacements, velocities, accelerations, crack growth, corrosion progress, changes in the chemical composition of building materials, etc. Because of the low cost of computer and communications technology, it is feasible to monitor simultaneously several parameters in real time and configure the monitoring systems so that they trigger an alarm as soon as monitored parameters exceed predefined threshold levels. Often, however, a simple monitoring strategy, such as measuring the distance between critical points of a structure once or twice a year, is more than adequate to assess the overall deformational behavior of the structure.

REDUNDANT DESIGN

The Swiss engineer Heinz Isler (Ramm, 1986), an innovator in the area of shell structures, applies a design method based on redundancy (not to be confused with structural redundancy). As a first step, Isler does experiments to determine an efficient shell shape. Afterward, he carries out simple, manual calculations in order to estimate the load-carrying capacity of the selected shell shape. If the results are satisfactory, Isler measures the shape of the shell and uses it to carry out computer calculations. In addition, and if deemed necessary, Isler does experiments on scale models (*Fig. 4*) to verify the results of the computer programs. After the shell has been built, Isler may decide to monitor it for several years. In this manner, he can determine if the design assumptions were correct. Isler's approach is redundant and, therefore, very reliable. Probably, this is the reason why none of his 1,000 or more shells have failed.

Fig. 4. Model test in Isler laboratory

A design approach similar to the one applied by Isler is feasible for all kinds of structures, not just shells. Such a design approach, which I propose to call *redundant design,* consists of the following steps:

1. Preliminary design using simple, manual calculations. If available, using several calculations methods based on different simplifying assumptions
2. Computer calculations
3. Verification of the plausibility and validity of the computer calculations using simple, manual calculations
4. For complex structures: additional verification of the plausibility and validity of the computer calculations using several different computer programs. If available, using programs based on several different calculation methods such as finite, boundary, or hybrid element methods
5. Experiments on scale models to verify the results of the computer calculations
6. Monitoring to confirm, modify, or reject the design assumptions and obtain feedback on how to improve future designs
7. Feedback to one of the previous steps if differences between predicted and effective values are determined

SUMMARY AND CONCLUSION

This paper discusses the following methods with the objective of preventing construction failures: 1) statistical investigations of construction failures, 2) research on construction failure case studies, 3) transfer of failure prevention methods from risk-conscious industries to construction, 4) peer reviews, 5) monitoring, and 6) redundant design. *Fig. 5* shows the phases of the construction life-cycle during which the discussed methods should ideally be used.

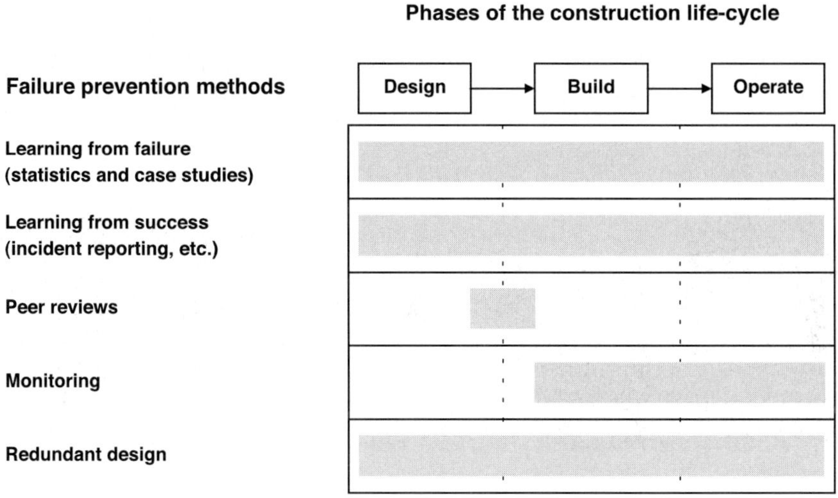

Phases of the construction life-cycle

Fig. 5. Phases of the construction life-cycle with corresponding failure prevention methods

The largest benefits in the prevention of construction failures are accomplished when several prevention methods are applied simultaneously and reinforce each other.

ACKNOWLEDGEMENT

The work on this paper was supported by the Basic Research Commission and the Institute for Technology Management, University of St. Gallen, Switzerland.

REFERENCES

Baur, H. (1981). *Untersuchungen über die bautechnische Prüfung und den Prüfingenieur für Baustatik.* Dissertation, Institut für Tragkonstruktionen, Universität Karlsruhe, Karlsruhe (in German).

Bell, G.R., Kan, F.W., and Wright, D.R. (1989). "Project Peer Review: Results of the Structural Failures II Conference," *Journal of Performance of Constructed Facilities,* Vol. 3, pp. 218-225.

Eldukair, Z.A. and Ayyub, B.M. (1991). "Analysis of Recent U.S. Structural and Construction Failures," *Journal of Performance of Constructed Facilities,* Vol. 5, pp. 57-73.

Faller, M. and Richner, P. (2000). "Sicherheitsrelevante Bauteile in Hallenbädern," *Schweizer Ingenieur und Architekt,* Nr. 16, April 20 2000, pp. 364-370 (in German).

Juran, J.M. and Gryna, F.M. (1993). *Quality Planning and Analysis: From Product Development through Use.* McGraw-Hill Inc., New York.

Matousek, M. and Schneider, J. (1976). *Untersuchungen zur Struktur des Sicherheitsproblems bei Bauwerken.* Birkhäuser, Basel (in German).
Pool, R. (1997). *Beyond Engineering: How Society Shapes Technology.* Oxford University Press, Oxford.
Ramm, E. (1986). "Form und Tragverhalten," *Heinz Isler: Schalen, Katalog zur Ausstellung,* Krämer, Stuttgart (in German).
Roberts, K.H. (1993). *New Challenges to Understanding Organizations.* Macmillan Publishing Company, New York.
Schaaf, van der, T.W. et al. (1991, eds.). *Near Miss Reporting as a Safety Tool,* Butterworth-Heinemann, Oxford.
Schneider, J. (1997). *Introduction to Safety and Reliability of Structures.* International Association for Bridge and Structural Engineering, Zurich.
Yin, R.K. (1994). *Case Study Research: Design and Methods.* Sage Publications, Thousand Oaks, California.
Yin, R.K. (1993). *Applications of Case Study Research.* Sage Publications, Newbury Park, California.
Zallen, R.M. (1990). "Proposal for Structural Design Peer Review," *Journal of Performance of Constructed Facilities,* Vol. 4, pp. 208-215.

Mitigation of failures due to inappropriate loading during construction – a European code

BRIAN S. NEALE, Chairman of the Project Team developing Eurocode: Actions during Execution; Member of the Standing Committee on Structural Safety; Technology Division, Health and Safety Executive, Bootle, UK

INTRODUCTION

The results of forensic engineering investigations reveal reasons for failures. These failures may range from serviceability [1] to catastrophic. The given reasons for failures in performance will usually centre on the technical aspects, but may also include results of investigations into the trails of actions or inactions that led to a particular failure. The management of the process can thus also come under scrutiny. The findings usually show that there is more than one reason for a failure, and usually many reasons [2], [3]. Such reasons can include inappropriate or inadequate management of projects, both during construction processes and when the facility has been completed. The reasons (or causes) can include:

1. poor quality of build, e.g. design, materials and workmanship;
2. inappropriate methods of build, including loadings imposed during the process;
3. delivering more efficient structures within a competitive environment, and perhaps novel features (which are inadequate in some way);
4. inadequate monitoring of existing facilities to deal with changing conditions, including loadings;
5. ineffective use and maintenance strategies by owners and/or controllers of facilities;
6. not realising, or understanding, the consequences of particular actions or inactions;
7. inadequate knowledge and experience of those involved;
8. ineffective management of knowledge, including lack of ready access to information.

The Standing Committee on Structural Safety which monitors trends in structural failures from its base in the UK, has recognized the variety of circumstances that can contribute to failures and publishes a report on a biennial basis of its work [4].

The construction processes, mentioned above, are thought of by many as relating to permanent works only, but temporary works are also relevant. Additionally, the term "construction processes2 can be limited in the minds of some to new build only. The range, however, is far wider from, for example, from green field build through to demolition, and can include refurbishment, partial demolition and rebuilding works, for example. The latter activities now tend to form a greater proportion of industry workload as more of the existing stock of structures and buildings are brought back to required serviceability states and perhaps upgraded and perhaps put to new uses, some where the existing materials are retained.

Forensic engineering: the investigation of failures. Thomas Telford, London, 2001.

Within this context, and in particular for 2 above, a European standard is being developed to give guidance [5] to help mitigation of failures due to inappropriate loading. This can include those such as overloading or inappropriate positioning of loads, at various stages during the development and use of a structure, and during the types of activities just described above. The standard, when published will be known as a Eurocode called EN 1991-1-6 "Actions on structures: Actions during execution". This means loadings to be taken into account on structures during the types of "construction" activities described above, and includes both permanent and temporary structures. The standard will be published by the Comité Européen de Normalisation (CEN), the European standards body for this type of work.

A tendency towards leaner, more efficient, more economical – and in some cases, innovative – structures has presented a greater need for a code on this topic. Such structural requirements as these, it can be argued, are what structural engineers should be considering as part of their professional approach. Indeed, the Institution of Structural Engineers, based in the UK, defines structural engineering as:

> "The science and art of designing and making, with economy and elegance, buildings, bridges, frameworks, and other similar structures so that they can safely resist the forces to which they may be subjected."

This paper describes the development of the Eurocode EN 1991-1-6, to the August 2001 pan-European consultation draft, known as the Stage 32 draft. It is described in the context of other Eurocodes, which includes a description of the structural engineering Eurocode system. A Code is also being developed independently in the USA [6], which has many similar aims.

STRUCTURAL EUROCODES

Background

As a general background, in 1975 the Commission of the European Community decided to embark on a programme for construction activities. The objective of the programme was the elimination of technical obstacles to trade and the harmonisation of technical specifications. Within this programme, the Commission decided to establish a set of harmonised technical rules for the design of construction works which, in a first stage, would serve as an alternative to the national rules in force in the Member States and, ultimately, would replace them. For fifteen years, the Commission, with the help of a Steering Committee with representatives of Member States, developed the programme, which led to the first generation of European codes in the 1980s.

In 1989, the Commission and the Member States of the European Union (EU) and European Free Trade Association) EFTA decided to transfer the preparation and the publication of the Eurocodes to CEN through a series of mandates, in order to provide them with the future status of European Standard (EN). This was on the basis of an agreement[1] between the Commission and CEN. Hence this links the Eurocodes with the provisions of all the Directives and/or Commission's decisions dealing with relevant European standards.

---

[1] Agreement between the Commission of the European Communities and the European Committee for Standardisation (CEN) concerning the work on EUROCODES for the design of building and civil engineering works (BC/CEN/03/89).

Status and application

The status and field of application of Eurocodes is such that member states of the EU and EFTA recognise that Eurocodes serve as reference documents for the following purposes:

- as a means to prove compliance of building and civil engineering works with essential requirements such as Essential Requirement N°1 – Mechanical resistance and stability, and Essential Requirement N°2 – Safety in case of fire

- as a basis for specifying contracts for construction works and related engineering services

- as a framework for drawing up harmonised technical specifications for construction products

Eurocode standards recognise the responsibility of regulatory authorities in each Member State and have safeguarded their right to determine values related to regulatory safety matters at national level where these continue to vary from State to State.

The National Standards implementing Eurocodes will comprise the full text of the Eurocode (including any annexes), as published by CEN, which may be preceded by a National title page and National foreword, and may be followed by a National annex.

The National annex may only contain information on those parameters which are left open in the Eurocode for national choice, known as Nationally Determined Parameters, to be used for the design of buildings and civil engineering works to be constructed in the country concerned. These include, for example:

- values for partial factors and/or classes where alternatives are given in the Eurocode;

- values to be used where a symbol only is given in the Eurocode;

- country specific data (geographical, climatic, etc). e.g. snow map;

- the procedure to be used where alternative procedures are given in the Eurocode;

- decisions on the application of informative annexes;

- references to non-contradictory complementary information to assist the user to apply the Eurocode.

This standard gives alternative procedures, values and recommendations for classes with notes indicating where national choices can be made. Therefore the national standard implementing EN 1991-1-6 should have a National annex containing all Nationally Determined Parameters to be used for the design of buildings and civil engineering works to be constructed in the relevant country.

The Eurocode system

The Eurocode standards provide common structural design rules for everyday use for the design of whole structures and component products of both a traditional and an innovative nature. Unusual forms of construction or design conditions are not specifically covered and additional expert consideration will be required by the designer in such cases. The Eurocodes are an integrated set of international codes of practice for the design of buildings and civil engineering works which will ultimately replace the differing rules in the various Member States of the EU and EFTA, although National Annexes will be permitted for some criteria, as described above. Appropriate levels of risk and reliability are built in to the codes [7].

The Structural Eurocode programme comprises the following standards, each one generally consisting of a number of Parts:

EN 1990	Eurocode	Basis of Structural Design
EN 1991	Eurocode 1:	Actions on structures
EN 1992	Eurocode 2:	Design of concrete structures
EN 1993	Eurocode 3:	Design of steel structures
EN 1994	Eurocode 4:	Design of composite steel and concrete structures
EN 1995	Eurocode 5:	Design of timber structures
EN 1996	Eurocode 6:	Design of masonry structures
EN 1997	Eurocode 7:	Geotechnical design
EN 1998	Eurocode 8:	Design of structures for earthquake resistance
EN 1999	Eurocode 9:	Design of aluminium structures

EN 1990: Basis of Design [8], the lead code, is independent of material and is an operational code of practice which deals with the principles and requirements for the safety, serviceability and durability of structures. Additionally it describes the basis for their design and verification as well as giving guidance for related aspects of structural reliability. In general terms, EN1992-EN1999 can be regarded as "resistance" material codes with EN1991 giving the actions (or loads) to be considered for resisting.

Criteria concerning safety, serviceability, robustness, for example, are deemed to be met only if the Principles and Application Rules are complied with. Additionally, various other conditions should be satisfied to give full effect to the desired outcome. These include adequate supervision, quality control during the execution of the works and subsequent suitable maintenance.

CONTEXT AND CONTENT OF ACTIONS DURING EXECUTION

Context

This European Standard EN 1991-1-6 "Actions on structures: Actions during execution" is relevant for all those concerned with the process including clients, designers, constructors and public authorities. It will form Part of a suite of EN 1991 documents, as thus:

EN 1991-1-2 Eurocode 1: Actions on structures Part 1-2: Fire actions
EN 1991-1-3 Eurocode 1: Actions on structures Part 1-3: General actions: Snow loads
EN 1991-1-4 Eurocode 1: Actions on structures Part 1-4: General actions: Wind actions
EN 1991-1-5 Eurocode 1: Actions on structures Part 1-5: General actions: Thermal actions

EN 1991-1-6 Eurocode 1: Actions on structures Part 1-6: Actions during execution
EN 1991-1-7 Eurocode1: Actions on structures Part 1-7: Accidental actions due to impact and explosions
EN 1991-2   Eurocode 1: Actions on structures Part 2: Traffic loads on bridges

The code EN 1991-1-6, as so far drafted, provides principles and general rules for the assessment of actions, climatic and environmental influences which should be considered, and designed into the execution stages of buildings and civil engineering works, including structural alterations such as refurbishment and rehabilitation, including partial or full demolition [9]. The term "construction" therefore should be seen the wider context of the industry as a whole in that it includes those more specialized activities.

Significantly, as well as covering "permanent" structures, it also covers "auxiliary" structures. These should be considered as non-permanent structures, as well as any works associated with the main activities which are to be removed after use (e.g. falsework, scaffolding, propping system, cofferdam, bracing, launching nose). However, it does not give design rules for formwork and falsework. These are being developed under a different series of CEN Codes by different teams. Those codes will, however, refer to EN 1991 for actions to be considered for the design of such structures when all the relevant codes are completed.

Actions during execution gives actions to be considered and does not give rules for the execution of buildings and civil engineering works.

Symbol notations used are based on ISO 3898 : 1987, supplemented as necessary in EN 1990 with additional notations specific to this Part, e.g.

F_h the horizontal force transmitted by a bridge deck built by the incremental launching method from the piers.

Structure of document

EN 1991-1-6 is structured in section and annexes as follows:

Section	Title
Section 1	General
Section 2	Classification of actions
Section 3	Design situations and limit states
Section 4	Representation of actions
Normative Annex A1	Supplementary rules for building
Normative Annex A2	Supplementary rules for bridges
Informative Annex B	Actions on structures during other transient design situations

Principles and application rules are given in the code for the assessment of actions and also for environmental influences that are to be consideration during execution stages of buildings and civil engineering works. These actions include:

- selfweight of structural and non-structural elements;
- actions caused by ground;
- prestressing;
- predeformations;
- temperature and shrinkage and hydration actions;
- wind actions;
- snow loads;
- water actions, debris effects;
- atmospheric ice loads;
- construction loads;
- (some) accidental actions;
- seismic actions.

Actions are to be classified, for example, for climatic and environmental influences as well as construction loads. Construction loads are those that can be present during the execution activities, but are not present on completion of the works. Construction loads are deemed to be those, which are caused, for example, by cranes, equipment, auxiliary structures and these are to be classified as fixed or free actions. Where construction loads are classified as fixed, tolerances for possible deviations from the theoretical position shall be defined. Where construction loads are classified as free, the limits of the area where they may be moved or positioned shall be determined.

Contents

Parameters are given for the identification and definition of particular design situations, including serviceability and ultimate limit states. During execution, only transient, accidental or seismic design situations are recommended for consideration, as appropriate.

The design situations should be selected for the structure as a whole, the structural elements and also for auxiliary structures and equipment, if relevant. For the various execution stages, the design situations should be identified taking into account conditions that apply from stage to stage in accordance with the execution process defined in the design and the methods of construction. Where deviations from the intended sequence of the execution process can occur, the design situations are to be reassessed.

Such conditions include the:

- support conditions;
- structural system;
- shape of the structure;
- degree of completeness, including the non-structural elements.

The effects and duration of transient activities is considered, as are the imperfections in the geometry of the structure and of structural members. These imperfections may be defined for the particular project. Dynamic effects and water actions are also considered.

Serviceability limit states are considered and advice given that operations should be avoided that can cause excessive cracking and/or early deflection during execution and which may affect the durability, fitness for use and/or aesthetic appearance in the final stage. Additionally, instead of taking account of the load effects due to shrinkage and temperature by means of design, appropriate detailing provisions may be selected to minimise these load effects. Within this context serviceability criteria for execution stages should be agreed for each project, and where this does not happen the serviceability limit states for the execution stage should be identical with those for the completed structure.

Serviceability requirements for temporary and auxiliary structures are to be defined in order to avoid any deformation and displacement which could affect the effective use or appearance of the structure, or cause damage to finishes or non-structural elements.

Ultimate limit states are expected to be verified for all transient and accidental design situations, including combinations of actions according to criteria in the Basis of Design document EN 1990. In this context, accidental design situations generally refer to exceptional conditions or exposure, such as impact, local failure and subsequent progressive collapse, fall of structural or non-structural parts, and, in the case of buildings, abnormal concentrations of building equipment and/or building materials, water accumulation on steel roofs, fire, for example.

Supplementary guidance is given for buildings (Annex A1) and for bridges (Annex A2). Actions on structures during other transient design situations are given (Annex B) where these include partial and full demolition

Construction loads (Q_c)

Construction loads need to be described as they could mean different things to different people and in the document they are seen to include:

- working personnel, staff and visitors, with small site equipment (Q_{ca});
- storage of movable items (e.g. building and construction materials, precast elements, and equipment) (Q_{cb});
- heavy equipment in position for use (e.g. formwork panels, scaffolding, falsework, machinery, containers) or during movement (e.g. travelling forms, launching girders and nose, counterweight) (Q_{cc});
- heavy equipment in free movement (e.g. cranes, lifts, vehicles, lift truck, power installation, jacks, heavy lifting devices) (Q_{cd});
-- variable loads from parts of the structure before the final design takes effect (e.g. fresh concrete, loads due to the process of construction such as assemblage) (Q_{ce}).

Ways of modeling these loads are given, together with nominal values if specific data is not available. For example, construction loads due to working personnel, staff and visitors, with small site equipment (Q_{ca}) should be modelled as uniformly distributed loads and applied to obtain the most unfavourable effects. The recommended value of Q_{ca} is given as equal to 1,0 kN/m^2, unless other specific information is known.

The horizontal effects of construction loads are to be taken into account in the structural design of a partly completed structure. Other horizontal forces to be applied which should be taken into account include wind forces and the effects of sway imperfections and sway deformations. Construction loads during casting of concrete are still under consideration, but are to be tied into considerations already made by others who are preparing the CEN codes for falsework and formwork.

Future development of the code

The programme for this code is to continue development, in consultation with the interested countries, leading to voting for implementation. Implementation is planned to be during 2003.

SUMMARY

The proposed Eurocode standard, when implemented, is designed to mitigate against some types of failures by giving rules and guidance on loadings and combinations of loadings to be considered for the planning and execution of works. It should also heighten awareness of the need to consider these parameters at an early stage and may then effect decision making for the proposed methods of work to provide greater project health and safety

ACKNOWLEDGEMENTS

The author thanks colleagues for their support in the preparation of this paper. The views being developed and expressed, however, are not necessarily those of CEN or the Health and Safety Executive.

REFERENCES
1. Neale B.S., The consequences of poor serviceability - and the way forward. *Management of Concrete Structures for long-term serviceability,* Thomas Telford Ltd, London, 1997.
2. Feld, Jacob and Carper, Kenneth L., *Construction Failure (Second edition),* John Wiley & Sons, Inc., New York, NY, 1997.
3. Neale B.S., "Forensic engineering in safety enforcement – some UK experiences", *ASCE Advancements in structural engineering through the analysis or failures - Structural Engineers World Conference, San Francisco,* American Society of Civil Engineers, Washington, 1999.
4. Standing Committee in Structural Safety, *Structural Safety 2000-01: Thirteenth Report of SCOSS,* SCOSS, London, 2001.
5. EN 1991-1-6 Eurocode 1: Actions on structures   Part 1-6: Actions during execution, Stage 32 draft, CEN, Brussels, 2001
6. American Society of Civil Engineers, *Design loads on structures during construction* (Draft 8/2000), ASCE, Reston, Virginia, 2000.
7. Calgarro,J-A and Gulvanessian,H, *Management of reliability and risk in the Eurocode system,* Proceedings of Reliability of Structures Conference in Malta, IABSE, Zurich, 2001.
8. Eurocode EN 1990: Basis of Design, Stage 34 draft, CEN, Brussels, 2001
9. British Standards Institution, *BS 6187:2000 Code of practice for demolition*, BSI, London, 2000.

APPENDIX A – Contents of EN 1991-1-6 Actions during execution

Contents

Foreword
Background of the Eurocode programme
Status and field of application of Eurocodes
National Standards implementing Eurocodes
Links between Eurocodes and harmonised technical specifications (ENs and ETAs) for products
Additional information specific for EN 1991-1-6
National annex for EN 1991-1-6

Section 1   General
1.1 Scope
1.2 Normative references
1.3 Distinction between principles and application rules
1.4 Definitions
1.5 Symbols

Section 2   Classification of actions

Section 3   Design situations and limit states
3.1 General – identification of design situations
3.2 Serviceability limit states
3.3 Ultimate limit states

Section 4   Representation of actions
4.1 Characteristic values
4.2 Selfweight of structural and non-structural elements
4.3 Actions caused by the ground
4.4 Prestressing
4.5 Predeformations
4.6 Temperature, shrinkage, hydration effects
4.7 Wind actions (Q_w)
4.8 Snow loads (Q_{Sn})
4.9 Water actions and debris effects
4.10 Atmospheric ice loads
4.11 Construction loads (Q_c)
4.12 Accidental actions
4.13 Seismic actions

Annex A1 (normative)
Supplementary rules for buildings
A1.1 Ultimate limit states
A1.2 Serviceability limit states

Annex A2 (normative)
Supplementary rules for bridges
A2.1 Ultimate limit states
A2.2 Predeformations
A2.3 Snow
A2.4 Construction loads

Annex B (informative)
Actions on structures during other transient design situations

Development of the ASCE 37 Construction Loads Standard To Mitigate Construction Failures

JOHN F. DUNTEMANN, P.E., S.E., Senior Consultant, Wiss, Janney, Elstner Associates, Inc., Northbrook, Illinois, USA, and
DR. ROBERT T. RATAY, P.E., Consulting Engineer, Manhasset, New York, USA

INTRODUCTION

More failures of structures occur during construction than after the completion of projects. Advances in construction technology, new materials, more refined design methods with less margin for error, the construction of more daring structures, as well as the pressure of time and cost of financing appear to contribute to the proliferation of failures during construction. In building construction the situation is chronic not only in the United States but in other countries as well. The proximate causes of the failures appear to be non-adherence to good practices, break-down of organization, management and communication in the field, and the lack of a comprehensive and definitive design standard in the office that deals with performance criteria, temporary loads, strengths and stability during construction. Designers, contractors, building officials, and in particular the US Department of Labor, Office of Safety and Health Administration (OSHA), are paying more attention to the problem.

Standards by themselves will not eliminate construction failures. They are not a substitute for experience, good judgment and care, but they will provide minimum criteria for safety and desired performance.

Current design codes and standards are mostly silent on the subject of construction loads, or give such general statements as "Proper provisions shall be made for stresses . . . during erection . . . of the building" and "Adequate temporary bracing shall be provided to resist wind loading . . . during the erection and construction phases." The questions, of course, are: what is adequate and what are proper provisions? The answers often depend on who gives them: the designer, the contractor, the owner, or the building official. They have little industry-wide guidance, make assumptions and judgments, and deal with consequent disputes.

To be sure, there are isolated manuals, guides and other forms of information put out by federal and state government agencies, public authorities, and industry organizations [Duntemann & Ratay; Handbook; Forensic]. Some foreign countries, the UK among them, are doing better than the US in providing this type of information. There is a definite need in the US to establish unified design criteria, loads, load combinations and load factors for the design and checking of structures during their transient construction stages and of temporary structures that are used as support, access, and protection during construction.

BACKGROUND

At the February 1987 meeting of the then ANSI A58.1, now ASCE 7, "Minimum Design Loads for Buildings and Other Structures" Standards Committee, Robert Ratay proposed that a section on loads during construction be added. The attendees agreed that there was a need for such a standard but the general opinion was that it should not be part of ASCE 7, but rather should be a separate document. Representatives of model code (national building code) groups pointed out that the model codes dealt with completed buildings and did not address construction, and that inclusion of construction loads in ASCE 7 would therefore create problems for them in referencing this standard.

Members of several structural design and construction related organizations and committees, and practicing engineers were surveyed; they all enthusiastically endorsed the establishment of an ASCE committee to develop a new consensus standard or guideline.

In April 1987, Ratay submitted a proposal to ASCE's Technical Council on Codes and Standards (Now Codes and Standards Activities Division, CSAD, of the Structural Engineering Institute, SEI of ASCE) to form a standards committee to embark on the new standards activity [Ratay]. After six months of reviews and deliberations, in October 1997 the Society's Board of Directors approved the new activity aimed at developing an ASCE Standard for *Design Loads on Structures During Construction*, and authorized the formation of a committee.

A meeting of twenty-five construction industry officials was convened in May 1988 to outline the general direction of the proposed standard. Several key ASCE participants met in September 1988 to begin the organization of the Standards Committee and to issue a call-for-members. The response by members and non-members of ASCE was overwhelming: a committee of ninety-six people was formed. The Committee, through its six subcommittees, has been working on the development of the standard since the spring of 1989. All procedures have followed approved American National Standard Institute (ANSI) rules for consensus standards. Several drafts of the proposed standard have been developed by small expert groups of members and issued for full committee balloting in 1996, 1997, 1998, and twice in 1999. The "affirm with comment" and "negative with comment" votes returned by the members were reviewed, and their proposed resolutions were developed by the appropriate subcommittees after each ballot. A revised draft was then prepared and issued for another round of balloting, until in late 1999 all issues within the committee were resolved. In October, 2000 the 10/2000 Draft was advertised and issued for public review and comment. One hundred thirty-six organizations and individuals requested and received the document, and many, though not all, returned comments in the form of votes with explanations. Resolutions of the "affirm with comment" and "negative with comment" votes received from committee members and the public were then proposed by the subcommittees, and discussed at the May 2001 meeting of the committee (in which several non-committee and non-ASCE members participated). The resolutions that were agreed to were incorporated into the 8/2001 Draft [ASCE 37] and issued in August 2001 for ballot to the full committee and for comments by those who returned public comments in the previous round. At the time of this writing, the ballots and comments on the 8/2001 Draft are being reviewed and publication of the final document is anticipated in early 2002.

PROPOSED STANDARD PROVISIONS

The objective of the new standard, *ASCE 37, Design Loads on Structures During Construction,* is to establish design loads, load combinations and safety factors to be used in the analysis and design of structures during their transient stages of construction, as well as temporary structures used in construction operations. The intent is that "The design loads shall provide for a level of safety of partially completed structures, and temporary structures used in construction, that is comparable to the level of safety of completed structures."

The standard is composed of six sections: a general introduction identifying the purpose and scope of the document; loads and load combinations; dead and live loads; construction loads; lateral earth pressures; and environmental loads. The standard includes allowable stress design (ASD) and load and resistance factor design (LRFD) provisions. The Table of Contents, indicating the breakdown of the sections, is included at the end of this paper.

The standard is presented in two-column format: the "standard" on the left side of the page in mandatory language, and the parallel "commentary" on the right side in explanatory language.

Construction Loads and Load Combinations

The construction loads, load combinations, and load factors were developed to account for the relatively short duration of load, variability of loading, variation in material strength, and the recognition that many elements of the completed structure that are relied upon implicitly to provide strength, stiffness, stability, or continuity are not present during construction. The load factors are based on a combination of probabilistic analysis and expert opinion [Rosowsky]. The concept of using maximum and arbitrary point-in-time (APT) loads and corresponding load factors is adopted to be consistent with ASCE 7 [ASCE 7].

The specification of particular load categories was made to facilitate logical load combinations to accommodate a variety of realistic loading situations that occur during construction. A brief summary of the load categories is as follows:

Dead load (D): total vertical weight of all completed permanent construction.

Live load (L): loads produced by the planned occupancy of the completed portions of the structures.

Fixed dead load (C_{FDL}): construction material loads, which are assumed constant during a certain phase or all of the construction period.

Variable dead load (C_{VDL}): construction material loads, which vary in magnitude during the construction period.

Worker and light equipment loads (C_W): loads due to workers and light equipment, such as tool boxes, and the like.

Lateral pressure of concrete (C_C): loads resulting from concrete pressures on formwork during moving, pouring, pumping, or placing.

Lateral earth pressures (C_{EH}): horizontal load effects resulting from soil pressures.

Horizontal construction loads (C_H): any horizontal loads arising from worker and/or equipment operations.

Erection and fitting forces (C_F): forces resulting from erection of equipment including alignment, fitting, bolting, bracing, guying, etc.

Equipment reactions (C_R): reactions from heavy equipment (rated or nonrated).

Currently, very little statistical information is available on which to base the selection of construction load factors (in strength design). As a result, initial selection of load factor for this construction load standard was based on factors in ASCE 7. Adjustments to these factors were then made based on an understanding of the nature, both physical and statistical, of these loads. As with ASCE 7, load factors are selected largely to account for uncertainty in the actual loads. The greater the uncertainty, either due to statistical variability or insufficient data, the larger the load factor (for an assumed, independent additive load). Issues of referenced period, correlated loads, and mutually exclusive loads are also considered in establishing load factors.

A summary of the construction load factors for strength design is presented in Table 1. For each type of load, a factor for the maximum load value is indicated and, where applicable, an arbitrary point-in-time (APT) load factor is also shown. There are a number of loads for which no APT load factor is provided; they should only be considered in load combinations when they are actually present and are therefore at their full or maximum value. Table 1 also indicates factors which, at the time of this writing, remain to be developed or may be selected from additional analysis. Certain cases also exist where there may be more than one possible load factor for maximum load values. This is indicated in Table 1 for load factors for lateral pressure of concrete (C_C) and for heavy equipment reactions (C_R). In the case of lateral pressure of concrete, a lower factor is specified for conditions of full fluid head since that presents less uncertainty than the condition of partial (unknown) fluid head. Similarly, a lower load factor is specified for heavy equipment reactions when the maximum equipment loads are rated by the manufacturer or are otherwise known with certainty. In both of these cases, no APT load factors are provided, since it is assumed that these are special conditions which must be considered in design only when the loads are known to be present.

Table 1. Construction Load Factors

Load	Description	Factor for maximum value (Y_{max})	Factor for arbitrary point-in-time value (Y_{APT})
D	Final dead load	1.2	--
L	Final occupancy live load	1.6	1.0
C_{FDL}	Fixed construction dead load	1.2	--
C_{VDL}	Variable construction dead load	1.4	By analysis
C_W	Worker and equipment load	1.6	0.5
C_C	Lateral pressure of concrete	1.3 (full head)	--
		1.5 (others)	--
C_{EH}	Lateral earth pressures	1.6	--
C_H	Horizontal construction loads	1.6	0.5
C_F	Erection and fitting forces	2.0	By analysis
C_R	Heavy equipment reactions	2.0 (unrated)	--

Environmental Loads and Load Factors

The following environmental loads are considered in the Standard:

> Wind load *(W)*
> Thermal load *(T)*
> Snow load *(S)*
> Earthquake load *(E)*
> Rain load *(R)*
> Ice load *(I)*

The basic reference for the computation of environmental loads is also ASCE 7. However, modification factors have been adopted to account for reduced exposure periods. For example, the design wind speed is the basic wind speed in ASCE 7 modified by the following duration factors for the period of exposure:

Construction/Exposure Period	Factor on design wind speed
Less than six weeks	0.75
From six weeks to one year	0.8
From one to two years	0.85
From two to five years	0.9

Furthermore, certain loads may be disregarded due to the relatively short reference period associated with typical construction projects, and certain loads in combinations may effectively be ignored because of the common practice of shutting down job sites during these events, e.g., at times of excessive snow and wind.

The load factors for the environmental loads are shown in Table 2.

Table 2. Environmental Load Factors

Load	Description	Factor for maximum value (Y_{max})	Factor for arbitrary point-in-time value (Y_{APT})
W	Wind load	1.4	0.5
T	Thermal load	1.4	--
S	Snow load	1.6	0.5
E	Earthquake load	1.0	--
R	Rain load	1.6	--
I	Ice load	1.6	--

ASCE 7 specifies an importance factor that adjusts the basic loads upward or downward depending upon the end-use occupancy and on the consequences of failure of the structure. (Critical structures, such as emergency facilities and places of assembly, are designed for greater loads than are most office buildings; other structrues, such as agricultural buildings that have low human occupancy, can be designed for lower loads.) For this construction loads standard, the importance factor is assigned a value of 1.0 for all structures, regardless of their end-use occupancy. This is consistent with the level of safety this standard intends to provide.

SUMMARY

The proposed ASCE standard provides a rational method for determining design loads on structures during construction subject to shorter than long-term exposure periods. Careful evaluation is needed to determine what impact this standard will have on current practice. While the authors recognize that standards alone will not eliminate construction failures, this standard will provide minimum criteria for safety and performance, which should mitigate the occurrence of construction failures.

REFERENCES

Duntemann, J.F., and Ratay, R.T., "Review of Selected U.S. and Foreign Design Specifications for Temporary Works-Part I," *Proceedings of the ASCE Structures Congress '97*, Portland, Oregon, 1997, pp. 985-990.

Handbook of Temporary Structures in Construction, 2nd ed., R.T. Ratay, McGraw-Hill, Inc., New York, 1996; Chapter 4, "Codes, Standards and Regulations," by J.F. Duntemann

Forensic Structural Engineering Handbook, R.T. Ratay, Mc Graw-Hill, Inc., New York, 2000; Chapter 2, "Design Codes, Standards and Manuals," by J.F. Duntemann

Ratay, R.T., "Standards for Design Loads During Construction: An ASCE Effort," *Proceedings of the ASCE Structures Congress '89*, San Francisco, California, 1989, pp. 870-875.

ASCE 37, "Design Loads on Structures During Construction," Draft 8/2001, American Society of Civil Engineers, Reston, Virginia, 2001.

Rosowsky, D.V., "Load Combinations and Load Factors for Construction", *ASCE Journal of Performance of Constructed Facilities,* November 1996, Vol. 10, No. 4, pp.175-181.

ASCE 7-95, "Minimum Design Loads for Buildings and Other Structures," American Society of Civil Engineers, Reston, Virginia, 1995.

DESIGN LOADS ON STRUCTURES DURING CONSTRUCTION

Case study: design, construction, and serviceability issues for a precast concrete facade subject to hurricane-force wind loads

Mark K. Schmidt, Consultant, Petar Plemic, Consultant, and Ian Chin, Principal
Wiss, Janney, Elstner Associates, Inc., Northbrook, Illinois, U.S.A.

ABSTRACT

It is the authors' intent to demonstrate the variety of issues to be considered and the proper repair techniques to be implemented on a precast concrete facade that suffered from a variety of afflictions. The facade repairs for the subject building addressed design and construction deficiencies including panel cracking, corrosion related distress at anchorages, and installation of supplemental panel fixings (supports) to resist hurricane-force wind loads. This paper also describes the selection of joint sealant and a unique two component clear water repellent system, which was used to preclude further streaking and eventual etching of the glass surfaces. Lessons learned from this case history could be used to establish general guidelines for evaluation, repair and protection of facades of similar type.

BACKGROUND

The subject structure is a thirty-story reinforced concrete frame building built in 1988 and located in the southeastern United States. Exterior facade elements above the third floor include precast concrete panels, structurally glazed punched windows and aluminum louvers. There are approximately 1200 precast concrete panels comprising about 450 different types on the building. The precast concrete panels are nominally 5 inches thick, with an exposed aggregate concrete face mix and a conventional concrete mix for the reminder of the panel. Most of the panels are rectangular with one, two or three openings for windows or louvers. The exterior glass surfaces are flush with the exterior precast panel surfaces. Some rectangular panels contain no openings. The panels are reinforced with one layer of 4x4-W4.0xW4.0 welded steel wire fabric and No. 4 (13 mm) steel reinforcing bars. The bars are typically used at panel perimeters, around window openings, and at lifting inserts. The panels are connected to the building frame by steel angles or plates that are welded to embedments in the panels and slab edges.

The facade cladding on lower floors consists of a nominal ¾-inch (19 mm) thick granite panel in lieu of the exterior exposed aggregate concrete matrix. The granite panels are anchored to a backup concrete panel with ³/₈-inch (10 mm) diameter rods. A polyethylene slip sheet separates the granite cladding and concrete backup.

During a prepurchase inspection of the partially completed structure, concerns were raised regarding significant through-thickness cracks, no provision for movement at the welded panel connections, and potential panel bending strength deficiencies.

Forensic engineering: the investigation of failures. Thomas Telford, London, 2001.

INVESTIGATION
Interior examination of panels

A survey of the panels revealed cracking mostly in panels with openings for windows and louvers. The cracks were typically horizontal or diagonal in orientation as shown in Figure 1.

Figure 1. Typical horizontal and diagonal cracks viewed from interior

Observations of the steel embedments in the floor slabs revealed no signs of distress. However, many of the welded connections between the panel and slab embedments were constructed contrary to the original shop drawings. Visual observations confirmed that the slab embedments had not been cast flush with the top of the floor slabs as designed. There may also have been problems with panel alignment or construction or fabrication tolerances.

The as-built detail most frequently used at the top of the panel consisted of a single round or flat bar stock welded between the slab embedment and the panel embedment. Where this modification could not be implemented, supplemental wedges were welded between the upper panel embedment and the angle connecting the lower panel embedment to slab embedment (Figure 2).

Figure 2. As-built connection

Exterior facade survey
Binocular, as well as close-up and concrete delamination surveys from suspended scaffolding platforms, were performed. Cracked, delaminated and spalled concrete was evident at locations of panel anchorage embedments and lifting inserts that were used to strip the panels from the forms. The worst conditions were typically observed for the panels at building corners and along the panel edges (Figure 3); at these locations the concrete cover was reduced due to a 1-inch (25 mm) chamfer.

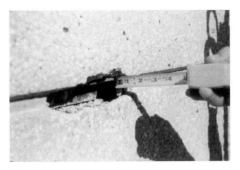

Figure 3. Deteriorated concrete along edges

Some cracks near embedments were narrow, typically vertical or diagonal in orientation, and are most likely the result of restraint to shrinkage and thermal movements. Now that these cracks have developed, most of the stresses that led to their formation have been relieved.

Wider cracks noted in the narrow portions of panels coincided with cracks observed from the interior (see Figure 1). These cracks occurred during fabrication as evidenced by rounded crack edges resulting from sandblasting to expose the aggregate at the panel surface (Figure 4).

Figure 4. Example of typical exterior crack

Some distress, in the form of cracks and displaced edges, was noted in the granite panel cladding. Since no shop drawings were available detailing the connection between the granite and the precast concrete, limited destructive exploration was conducted. As shown in Figure 5, the threaded rods used to fix the granite to the concrete backup were set in epoxy in angled holes with minimum granite embedments (measured from the inside granite surface to the nearest end of the rod) ranging from $^1/_8$ (3 mm) to $^5/_8$ (16 mm) inches. The limited embedments of the threaded rods raised concerns in light of the hurricane-force wind loads that may be imposed on the facade.

Figure 5. Threaded rod anchorage of granite cladding

Other observations included significant glass staining and possible etching (Figure 6). The joint sealant exhibited signs of significant weathering, common for its age and exposure, and poor installation.

Figure 6. Glass staining and etching

Panel reinforcement survey

Selected panels were surveyed from the interior with a metal detector. Findings including position and depth of reinforcement were verified by concrete chipping. On average, the center of the steel bars were 1½ inches (38 mm) from the outside face of the panels which is contrary to the specification requirement that no metal be within 2 inches (51 mm) of the finished concrete surface. From these measurements, it appears that the reinforcement was placed directly on top of the 1 to 1½ inches (38 mm) thick face mix during fabrication.

For the code-mandated wind load requirements in effect during the design of this project [1], the ideal placement of the reinforcement for the typical panel would have been approximately 2½ inches (64 mm) from the outside face (centered within the panel). The specification requirement that no metal be within 2 inches (51 mm) of the finished concrete surface was apparently an attempt by the designers to obtain the ideal placement of reinforcement. The deviation of the steel reinforcement from the specified location resulted in a reduction in load-carrying capacity under positive wind pressures.

STRUCTURAL ANALYSES
Precast concrete panels
Applicable sections of the code [1] were used to determine required wind loads and allowable stresses. In analyzing the panels, several factors were considered: specified concrete and steel strengths, as-built sections and properties, modification of the ACI [2] strength reduction factor, eccentricity of dead load, and an increase in allowable stress to account for load combinations including wind. The analyses were performed assuming cracked concrete sections, which accounted for any existing cracks as well as any future cracks that might develop.

A total of 363 panels above the 12th floor were found to be overstressed in bending based on both the 1984 [1] and 1994 [3] codes. The placement of the steel reinforcement was the primary factor in the calculated deficiencies. An insufficient amount of steel reinforcement was also a contributing factor.

Panel connections
The fully welded connections did not provide the degree of movement capability recommended by the Precast/Prestressed Concrete Institute (PCI). However, calculations performed on as-built connections revealed sufficient ductility to accommodate thermal or wind load-induced movements. The relatively good historical performance of all the connections and embedments indicated past adequate accommodation of movements resulting from shrinkage, creep, and thermal volume changes.

The modified connection consisting of supplemental steel wedges (shown in Figure 3) accounted for approximately 25 percent of the connections closely examined and had significantly less calculated ductility. The rigidity of these connections might cause cracking in the concrete adjacent to panel embedments under near-design wind loads, particularly as a result of interstory drifting. Because the welded wire fabric and reinforcing bars within the concrete panels bridge these cracks, this condition does not pose a structural concern. However, future cracks might lead to water leakage, associated corrosion of steel embedments and subsequent spalling of concrete. Following a major wind load event, the developed cracks might have to be injected with a flexible grout to minimize water infiltration. The resulting appearance might be aesthetically unacceptable, in which case an elastomeric coating could be applied over the precast concrete panels.

TESTING
Limited tests were conducted on the anchorages of the granite panel cladding. Prototypes of the minimum threaded rod embedment in the granite were constructed and tested to obtain an estimate of the fixing capacity based on a pullout failure of the rod embedded in the granite.

The average ultimate pullout capacity was approximately 140 pounds (623 N). Using a factor of safety of 4, the allowable capacity based on the tested failure mode was approximately 35 pounds (156 N). The wind load demands per the 1984 code were nearly twice the allowable calculated capacity; the 1994 code yields even higher wind load demands. In addition, the limited testing revealed that pullout capacities at locations with large quantities of biotite crystals could be less than half of the calculated average ultimate capacity. As a result, supplemental fixings for the granite cladding were warranted.

REPAIR
The repairs required to address current distress and to provide long-term serviceability of the building facade included modification of overstressed panels, repair of delaminated concrete due to corrosion of embedded steel, protection of the exposed concrete surfaces by the

application of an appropriate water repellent system, and replacement of the deteriorated sealant system.

Panel bending strength deficiencies
For panels with calculated bending strength deficiencies, hundreds of additional analyses were performed to determine the effect and required strength of supplemental fixings. These fixings, typically installed between the inside face of the panel and bottom of the concrete spandrel beams, effectively reduced bending moments by transforming the panels from single to multiple span members.

Supplemental fixing design criteria included ductility to accommodate in-place panel movements and ease of installation, considering limited access and heavily reinforced spandrel beam soffits. The resulting designed connection, shown in Figure 7, consisted of steel angles fastened with a prescribed number of concrete screws. The concrete screws provided the required strength without the need for deep embedment. By intentionally creating a gap between the steel angle and the panel, the screws also provided the required ductility.

Figure 7. Supplemental supports

Corrosion-induced distress
Typical concrete repairs were performed by sawcutting the perimeter of the panel repair area and removing all of the distressed concrete to expose the corroded steel. The exposed concrete and steel surfaces were then sandblasted to remove corrosion by-products and to clean the surfaces. After application of a protective coating to the exposed steel surfaces, the concrete section was restored by the placement of new concrete. After curing, the concrete patch perimeter was ground and sealant was installed.

In order to economize and maintain aesthetics, the recommended repairs to the concrete at embedded steel angles and lifting inserts were different below the 11th floor. The lower floor repairs did not receive perimeter sealant in order to achieve a more finished appearance at higher visibility areas.

Granite cladding anchorage deficiencies
Supplemental fixings consisting of stainless steel concrete screws in countersunk holes were specified to remedy the calculated deficiencies under the design wind loads. The number and pattern of anchors required varied depending on the granite panel size and its proximity to discontinuities such as building corners. The 11 mm countersunk holes created in the granite were filled with a complementary colored sealant. A trial repair, shown in Figure 8, was

installed on the building and was found to provide a readily feasible and aesthetically acceptable solution. At this time, widespread anchorage repairs to the granite cladding have not been completed.

Figure 8. Trial installation of supplemental granite panel fixing

Serviceability repairs

All existing polyurethane sealant and backer rod between precast concrete panels were removed and replaced with new backer rod and silicone sealant. Silicone was recommended as the replacement sealant since its anticipated service life is longer than that of polyurethane sealants. However, the use of silicone sealant in this application presented several concerns.

First, silicones have a greater tendency than polyurethanes to attract and hold dirt. Secondly, components in the silicone sealant might stain the precast concrete adjacent to joints. Staining could take the form of an increased attraction to airborne dirt or a continuous wet appearance near the joint. During periods of precipitation, staining may also appear as a light-colored waterproof zone adjacent to joints. Once stains develop, they are difficult, if not impossible, to eliminate.

Precast concrete panels are not one of the more susceptible substrates with respect to staining. In order to minimize the potential for staining and attraction to dirt, a specially formulated silicone sealant with a relatively stain-free history and a matte-type finish was selected. In addition, preconstruction substrate stain testing was specified.

A water repellent system applied to exterior surface of the exposed aggregate precast concrete panels was recommended for two reasons: it would inhibit future deterioration of the precast concrete panels and it would minimize the potential for further streaking and etching of the window glass. A two-component water repellent system (consisting of a penetrating sealer and a clear acrylic topcoat) was subsequently applied to minimize the potential of future glass etching caused by alkaline substances leached from the concrete by rainwater.

LESSONS LEARNED

This building suffered from most of the problems that could afflict precast concrete facades. Most of these problems can be prevented during design, panel fabrication, and erection of such facades. Proper design should address wind-induced stresses, thermal movements, and differential movements between the building frame and facade.

The precast concrete connection design should allow for movement but also for variation due to erection tolerances and eccentricities. Thin granite cladding mechanical anchorages must be designed and installed considering the required wind loads and limited embedment available. Where possible, glass should be recessed with respect to precast concrete panels or the concrete should be treated with an appropriate water repellent system.

The as-built structural frame dimensions should be verified prior to beginning fabrication of the precast panels. Accurate placement of the reinforcing steel is also critical, especially in higher wind zones.

6. REFERENCES

[1] *The South Florida Building Code, 1984* Edition.
[2] *Building Code Requirements for Reinforced Concrete* (ACI 318-83).
[3] *The South Florida Building Code,* 1994 Edition.
[4] *PCI Design Handbook,* 4th Edition.

The application of logic to evidence

DR. JOHN MCCULLOUGH
Cadogan Consultants, Glasgow, UK

ABSTRACT
Many cases proceed to litigation or arbitration on the basis of engineering evidence which is fatally flawed. There are simple methods for testing validity and reasoning which should be deployed early in the proceedings.

Keywords = reasoning, evidence, logic.

INTRODUCTION
This paper is based on two very similar publications I presented some years ago and which are listed in the bibliography.

There are many examples of cases of litigation or arbitration which proceed to a hearing or trial on the basis of evidence which is fatally flawed. Such cases should fail.

When such cases arise the ultimate clients are involved in the unnecessary expense of lawyers' and experts' fees to fight actions which if tested at the outset would have been shown to have been manifestly unsustainable.

Technological experts are usually not trained in logic. Lawyers even if not trained are generally better than technologists and engineers when it comes to logic. However, it often occurs that the plethora of technological evidence obscures the very essence of the case being advanced and spurious arguments go undetected or become specious.

There are some simple methods which can be applied to test the validity of an argument once it has been reduced to its skeletal or basic form[1][2]. If these methods were applied more generally in litigation and arbitration a great deal of nervous energy and money might be saved.

There are other fundamental principles of reasoning which relate to knowledge, definitions and fallacies and of which those involved in litigation or arbitration should be aware[1][2][3].

The purpose of this paper is to outline some of the twisted thought processes which can arise in the preparation of evidence and to describe some of the methods and considerations available for detecting them[1][2]. The methods for detection and analysis are very pedestrian and it is not suggested that they appear in evidential reports.

Forensic engineering: the investigation of failures. Thomas Telford, London, 2001.

However, they should be used to arrive at and check the more user-friendly information contained in reports.

EXAMPLE OF BOGUS ENGINEERING EVIDENCE

The writer has been involved in preparing evidence in a number of cases, large and small, which have involved weld and other engineering failures. The nature of the cases is actually irrelevant as far as the points to be made in this paper are concerned, which are used by way of example. The observations are applicable to any evidence or argument, regardless of subject.

If there is structural, pipeline or pressure vessel failure involving a poor weld, the following reasoning is often employed by those preparing evidence and advancing the case for the Claimant.

If the weld is poor it could cause failure.

There has been failure.

Therefore it has been caused by the poor weld.

Leaving aside for the time being such consideration as whether the weld was actually poor (this is a premise upon which the reasoning is based). The actual form of reasoning is fatally flawed and is known as affirming the consequent. It fails to consider and therefore discount competing probabilities.

In the case of the weld, competing probabilities could include unanticipated dynamic loading or a hostile environment, either or both of which would have caused a good weld to fail.

The invalidity of this form of reasoning is not so easy to detect when one is faced with prima facie impressive radiographic and metallurgical evidence of poor welding.

All of this impressive evidence must be set aside and the argument broken down to its skeletal form.

When the basic or skeletal form of the reasoning is identified and every day experiences are substituted, its bogus nature becomes apparent, viz.

If it is raining I will get wet.
I am wet.
Therefore it is raining.

This is obviously flawed because I could be wet for a number of reasons.

The writer has been involved in a number of cases where this form of flawed reasoning has formed the basis of one side's evidence and has propelled the case all the way to a hearing.

Each side should test the logic of its own case as well as that of the opponent. When testing the case of an opponent one must avoid assuming that the opponent's basic argument is too simple and the opponent's case should not be misrepresented. One must not commit the fallacy of arguing against a "straw man". One must always be charitable when trying to assess another person's form of argument. The skill lies in getting behind the evidence or pleadings to assess realistically where the other side is coming from in terms of thought processes.

LOGIC AND SYLLOGISMS
A recognised means of testing an argument is to break it down into syllogisms and to apply rules of logic to those syllogisms to test for validity.

A syllogism is a form of reasoning or deductive inference consisting of two premises and a conclusion. If one (or more) premise is not expressed it is known as an enthymeme.

There are different types of syllogism, e.g. hypothetical, conditional, disjunctive, categorical, etc.

The oldest form is the categorical which is also known as the classical or Aristotelian. It takes the following form.

Some temples are in ruins.
All ruins are fascinating.
Therefore some temples are fascinating.

This form of syllogism is generally not suitable for testing the logic of evidence. A more suitable form is the hypothetical syllogism and it is also easier to analyse. This paper will consider the hypothetical form only.

There are two valid forms of hypothetical syllogism: affirming the antecedent; and denying the consequent. There are two deceptively similar but invalid forms known as: denying the antecedent and affirming the consequent. In skeletal or algebraic form they are as follows[1][2].

1.   If p then q,
      p,
      therefore q   Valid

2.   If p then q,
      not q,
      therefore not p   Valid

3.   If p then q,
      not p,
      therefore not q   Invalid or Fallacious

4.   If p then q,
        q,
        therefore p   Invalid or Fallacious

The example referred to in the previous section is in the form of 4 above and is invalid because it affirms the consequent.

One has to cut through the plethora of technological evidence to see the argument in its skeletal form. Set aside all the radiographs, macrosections and metallurgical reports and take a look at what is actually being said, viz.

If the weld is poor (p) it could cause failure (q).
There was failure (q).
Therefore it was caused by the poor weld (p).

i.e.   If p then q,
        q,
        therefore p.   Invalid or Fallacious
                                                       (affirms the consequent)

If the radiographic and metallurgical evidence were such that the weld manifestly did not comply with normal specifications and therefore could prima facie be considered to be poor, the correct form of reasoning would be as follows.

If the weld is poor (p) then it could be a cause of failure (q).
The weld is poor (p).
Therefore it could be a cause of failure (q).

i.e.   If p then q,
        p,
        therefore q.   Valid
                                                       (affirms the antecedent)

This valid form merely identifies the poor weld as a possible cause of failure. Therefore other possible causes would be considered. A list of competing probabilities would be established and the most probable would be identified. The evidence and the case would proceed on the basis of logic and should be sustainable.

When using a syllogism which affirms the antecedent, or indeed generally when using hypothetical syllogisms, it is important to get the wording or meaning of the premises and conclusion to be consistent. The second premise should affirm the antecedent precisely as presented in the first premise and the conclusion must deliver the consequence precisely as presented in the first premise. Otherwise the fallacy of three terms may be committed[2]. In skeletal form the fallacy of three terms is as follows.

        If p then q,
        r,
        therefore q,

or
If p then q,
p,
therefore r.

When the invalid forms of syllogism (i.e. Nos. 3 and 4) are used competing probabilities are not considered and discounted. If the conclusion reached is correct then it is so by chance and not because of sustainable reasoning.

A distinction has to be drawn between valid argument and truth or proof. For a syllogism to establish proof the premises must be sound, e.g.

If my cat is a dog (p) then it is a book (q).
My cat is a dog (p).
Therefore it is a book (q).

i.e.   If p then q,
        p,
        therefore q.   Valid
                                                  (affirms the antecedent)

This argument is entirely valid but clearly does little to advance the accumulated wisdom of mankind. Its premises are absurd and truth or proof is not established.

Even if the premises are sound, there will be no proof unless a valid form of argument is used. It is not possible to demonstrate proof from sound premises using invalid argument.

THE LOGIC OF CAUSATION

In litigation or arbitration it is necessary to establish causation or liability and in this quest logic is also essential. The basic logic of causation is as follows[2].

To find a cause one must look for:

1.   another event C;
2.   that is prior to E, the event to be explained;
3.   is such that if C had not occurred E would not have occurred, all other things being equal;
4.   is such that if C had occurred in other similar circumstances, E would have occurred.

These four considerations are often not fully explored and taken into account when advancing evidence or allegations.

In a recent case it was alleged that poor welding was the cause of cracks that had grown in the welds. Several samples of welds, some allegedly poor and some allegedly good were submitted. Some had cracks and some had not, regardless of shape or appearance.

The illogicality of the claim that poor welding was the cause of cracking can be demonstrated in the following, albeit pedestrian, way.

If poor welding were the cause of crack growth (p) then poor welding would not be present in samples with no crack growth (q).
In fact poor welding was present in samples with no crack growth (not q).
Therefore poor welding was not the cause of crack growth (not p).

or

If crack growth were caused by poor welding (p) then cracks would not appear in good welds (q).
In fact cracks did appear in those welds described as good (not q).
Therefore crack growth was not caused by poor welding (not p).

In addition to complying with the logic of causation the above syllogisms are valid, viz.

> If p then q,
> not q,
> therefore not p.   Valid
>   (denies the consequent)

ADEQUACY OF KNOWLEDGE
This ancient philosophical principle has been well described by Schumacher[3].

As with aspects of logic it is obvious when one actually stops to think about it.

It may arise in litigation when for example a qualified and honest welder is asked to comment on a weld which has failed in a dynamic system in a corrosive environment or when trying to assess what a supplier should have appreciated at the outset.

The principle involved is that the knowledge of the observer must be fully adequate for the complexity of the subject, otherwise the subject cannot be fully understood.

An analogy of the following typical kind is used to explain the principle; it concerns a book.

(a)   If it is shown to a dog the dog will observe it to be a shape with a scent and possibly colour. The dog is correct for it is those things.

(b)   If it is shown to an untutored savage he will see in it more than the dog was able to. For example the savage might realise that it opens up and that the bits inside have patterns on them. The savage too is correct at his level of understanding.

(c)   If it is shown to an educated person but who is not knowledgeable of the language involved, he will realise that it is a book but will not know what it is about. That person too is correct.

(d)   If it is shown to a person educated in the same language then he will fully appreciate all that it is. That person too is correct.

The important point is that all four of the above observers are correct but only one has knowledge which is adequate for the full subject. The others are correct only insofar as they are able to go.

NARROWING AND BROADENING OF DEFINITIONS OR MEANINGS

This can occur in expert reports and witness statements. It may also occur during the giving of oral evidence and may be used by Counsel when examining witnesses.

It is usually easy to appreciate when narrowing or broadening is taking place because there are marker adjectives such as "true" or "real". It is often not so easy, at least not at first, to appreciate the full implications of these and similar adjectives.

They are rarely used gratuitously[1] in evidence or argument, unless the person using them is careless in the use of language and that would be unusual in such circumstances. The adjectives are used in an attempt to narrow or broaden a definition or meaning to exclude or include evidence to support the case of the arguer or witness.

One might expect to see their use in descriptions such as "true dynamic loading" or a "real coastal environment". The suggestion being offered is that the material type of dynamic loading is or is not dynamic loading or that the material marine environment is or is not as corrosive as a marine environment.

At a pollution trial Counsel invited the writer to acknowledge the difference between black smoke and true black smoke.

This approach of attempted narrowing or broadening is fundamentally unsound and unsustainable when detected. Also, it is a form of intellectual arrogance because implicit in it is the proposal that dictionary definitions are mere approximations and that by the application of pure thought a more appropriate meaning may be found[1].

FALLACIES

In addition to the fallacious syllogisms already referred to, there are a number of fallacial forms of argument which are well explained in books on logic[1][2]. These include the fallacies of division, composition, genetic, begging the question, Monte Carlo, bandwagon and memory transfer.

It is beyond the scope of this paper to try and list them all or to explain them. It is also beyond the writer's expertise as some of them involve psychology. However, some of those likely to be encountered in evidence will be outlined briefly.

The Monte Carlo fallacy is the belief that if a coin is tossed say 10 times and it comes up "heads" each time then on the 11th toss it is more likely to come up "tails". The reality of the situation is that with a normal or unbiased coin the chances are 50/50 whether the coin is tossed once or a million times.

The fallacy of begging the question arises when the premises depend upon the conclusion for validity. This flawed form of reasoning takes many forms, the following enthymeme is one of them[2].

This drug is soporific.
Therefore this drug will make me sleep.

The fallacy of composition is the belief that the whole will contain the properties or characteristics of the parts of which it is made up. The fallacy of division is exactly the opposite, that the parts will contain the properties of the whole.

The bandwagon fallacy is that which asserts that because a large number of people support or believe in a proposition it is correct.

The fallacy of novelty is the assertion that because something is new it is more likely to be correct. The fallacy of tradition argues exactly the opposite, i.e. because it has been done this way for years it is correct or good.

The fallacy of appealing to authority is often used in evidence and takes the form of asserting that something is correct because it is found in a book or is believed by an eminent person.

The fallacy of Ad Hominem Argument is sometimes committed by Counsel when attempting to discredit the facts or opinions as advanced by a witness. It introduces into an argument a premise about the person or witness advancing the fact or opinion. In essence it says that because a person has certain views or experience on one subject, or has self interest in the outcome, then evidence on a different or related subject will be unreliable. This simply cannot be sustained in logic.

CONCLUSIONS
Many actions are initiated and proceed to litigation or arbitration on the basis of crooked thinking. If the logic is put to the test at the trial or hearing then such cases should fail.

It would be better if those involved were to test the logic and thinking behind their own case as well as that of their opponent at an early stage in the proceedings. If this were done then great amounts of money and nervous energy might be saved.

REFERENCES
1. Shaw, P., Logic and its Limits, Pan Books 1981.

2. Luckhardt, C.G. and Bechtel W., How to do things with Logic, Lawrence. Elbroune Assoc., 1994.

3. Schumacher, E.F., A Guide for the Perplexed, Sphere Books, 1978.

BIBLIOGRAPHY

1.   McCullough, J, The application of logic to technological evidence in litigation, Journal of the International Society for Technology, Laward Insurance, Vol 2, No3, September 1997 E&FN Spon.

2.   McCullough, J. Logic before litigation, Journal of the Academy of Experts, Vol 3, No1, Spring 1998, FT Law & Tax.

Technically yes, but... Forensic engineering and insurance

EUR ING **RICHARD RADEVSKY** BSc CEng PEng FICE
MCIWEM MInstPet, MIFireE FCIArb MAE QDR,
Principal, Charles Taylor Consulting plc, London, UK (www.charlestaylorconsulting.com)

EXECUTIVE SUMMARY

Insurers and their advisors are major purchasers of forensic engineering services. Sometimes, but not always, the forensic engineers they use are given a clear brief. Often the brief has to be developed throughout the engineer's period of engagement. Forensic engineers may find themselves being pushed by Insurers and their representatives into addressing obscure hypothetical questions whose relevance is unclear. Then, when matters reach a crucial negotiating stage or come to court, the issues change rapidly, turning away from the area where most investigative effort has been concentrated.

For forensic engineers to provide the standard of service which will result in repeat instructions they need to understand their role in a wide context. The skill required to present evidence in a way that can be easily understood whilst remaining technically rigorous is almost as important as the ability to find the solution to a technical question. The forensic engineer needs to understand where his work fits in to insurance/legal processes to ensure that his client is being given the most appropriate advice. It is the variability in the quality of these latter skills that makes Insurers and their advisors cautious in their selection of forensic engineers.

INSURERS' NEED FOR FORENSIC ENGINEERING SERVICES

Insurers provide a variety of insurance products to cover construction, erection, professional liability, third party liability, employer's liability, personal accident, property and business interruption, machinery breakdown, operational risks and others. Claims involving any of these can require the services of a forensic engineer. Each is a different product providing different insurance cover with different conditions and exclusions. (Gloyn 1999)

When a problem arises it normally requires urgent attention. The precise nature of the problem may not be known immediately and so it is not obvious what type of engineering expertise is required, let alone who would be the ideal forensic engineer to appoint. In the early stages, Insurers may resist appointing any forensic engineer or they may rely on someone with general engineering expertise to make investigations to narrow down the issues. Such people are likely to be familiar with the work of a number of forensic engineers but none of these may have the precise expertise required even though some may have worked in the right field. A choice then has to be made either to go with someone who has a proven track record as a forensic engineer but possibly not in precisely the required specialism or to locate someone who works in exactly the right field but whose forensic engineering abilities are unproven.

There are of course specialist areas where Insurers regularly need forensic engineering expertise. Outside those areas where Insurers have established means of handling forensic

Forensic engineering: the investigation of failures. Thomas Telford, London, 2001.

engineering investigations, Insurers will start without a preconceived idea of who they should use. The challenge for Insurers is to avoid overloading these practices with work or setting up conflicts. This paper is not addressed at these areas. In some areas the forensic engineering expertise within the insurance and reinsurance community is probably the best that exists and there is relatively little need to use external expertise.

QUALITIES THAT INSURERS SEEK IN FORENSIC ENGINEERS
Specifically relevant expertise
What qualities does a good forensic engineer need? Firstly he/she must be an expert in the field. Unfortunately there is a large number of engineers who perhaps understandably will claim that their field of forensic engineering expertise is wider than their experience suggests. Rarely does one find that a forensic engineer will refuse an assignment because they do not feel that they have adequate expertise or because they know someone better suited than themselves. The consequence of this is that Insurers and loss adjusters have a deal of scepticism about forensic engineering abilities. If everyone says, "yes I am the person you need" then other means of determining suitability have to be employed.

The most common method of selecting a forensic engineering expert is recommendation; checking the experience of others. Past experience of a forensic engineer is obviously the most useful although two identical problems rarely occur. Affiliation to professional bodies that qualify experts (such as the Law Society) helps. Rarely would an insurer rely solely on databases of experts because such databases do not guarantee high quality.

Spoken/Written skills
The forensic engineer needs a collection of general skills in addition to his/her expertise. He/she needs to be able to write competently and perform well in court (should the matter develop into litigation). The ability to be technically rigorous can conflict with the need to be easily understood and to avoid jargon. A seemingly brilliant written report can be torn apart under cross-examination if the expert is unable to explain his/her reasoning clearly when under pressure.

Legal knowledge
The engineer needs knowledge of the law and contracts so that he/she can understand the context within which his/her investigation is set. Unfortunately the questions which are crucial to insurance claims are often not those that are technically the most interesting.

Case History 1
A leak from a gas pipeline was investigated to determine how a hole had formed in the pipe wall. An expert metallurgist was sent from UK to examine the pipeline hole and determine from the remains of the damaged pipe section what had caused the hole to form. Although completing this assignment successfully the forensic engineer then strayed into analysis of the structural engineering aspects on which he was not qualified to comment. His analysis was wasted and considerable time had to be spent limiting his report to the issue about which he had been asked to comment.

Experts frequently have to be restrained from pursuing technically fascinating investigations that have little bearing on the issues. Sometimes they try to cover more than the extent of their brief and this can pose dangers if they stray into areas where the level of their expertise is not sufficiently high.

FORENSIC ENGINEERING IN A NON-UK ENVIRONMENT

For historical and commercial reasons many major Insurers and Reinsurers are based in UK, Europe and North America. They tend to appoint forensic engineers from their own countries to investigate failures world-wide. When this happens it is vital that the forensic engineer realises that he/she is operating in a different physical, cultural, legal and professional environment. Climatic differences, differences in customs and practice, different legal structures, laws and codes of practice can markedly affect what conclusions should be drawn from an investigation.

Case History 2

A consulting engineer was responsible for designing and supervising the construction of an irrigation barrage in a South East Asian country. He told the contractor that he was very concerned about the soundness of a local contractor's temporary access bridge for heavy plant built across a large river. It was made from freshly felled tree trunks. Soon after completion, the barrage was completely destroyed following a piping failure in the soil beneath its foundations. Inadequate account had been taken of the soil conditions. The engineer surveying the scene of destruction was then faced with the contractor who quietly pointed out that his temporary tree trunk bridge (on which they were both standing) had lasted longer than the engineer's barrage.

Whilst the forensic engineer may be able to determine what has happened, Insurers are likely to need to know the significance of the engineer's conclusions set into the context of the country where the loss occurred. Insurers who cover risks in a particular country are expected to know about that country. Insurers would not normally be able to reject a claim on the basis of, say, a design that was defective by home country standards if it is not defective by the standards in the country of the risk. Forensic engineers need to be aware of this. Whilst Insurers may be interested in comparisons with a forensic engineer's home country, the outcome of a claim is unlikely to turn on such a comparison.

Case History 3

A Turkish building suffered from severe water ingress during heavy rainfall. A Turkish insurer who reinsured with an American reinsurance company insured the contents. A forensic engineer was engaged to determine whether the design of the roof was defective. Had this been the case the insurer of the goods may have been able to recover their loss from the building designers.

The forensic engineer was only able to compare the design of the roof drainage with British Standard requirements even though his company had a Turkish subsidiary. The conclusion that the roof did not meet British Standards was of no practical use and the analysis required to reach this conclusion wasted. As the building was Turkish owned and Turkish insured, Turkish law applied and in a court action a Turkish court would only have been interested in the Turkish Building Code. Judged against this the case for negligent design was considerably weaker.

INSURERS' OBLIGATIONS TO THEIR INSUREDS
Requirements of the Policy

Where large sums are involved, Insurers will not pay claims unless they have to. They will therefore look at a claim very closely, particularly to see if the cause of the loss is covered under the policy. Insurance policies impose conditions upon those insured and an insurer will want to see if the insured has breached a condition before deciding whether or not to pay a claim. Just identifying a breach is not, however, sufficient since the breach of the condition must be relevant to the loss. (I.e. had the condition not been breached would the loss not have happened?) Insurance policies come in a huge variety of forms with differences, which may be subtle, but nevertheless very significant. Forensic engineers need to know and understand subtleties such as the difference between foreseeable and foreseen. The success of a huge claim can turn upon such differences like these.

Preserving relationships where possible

The relationship between an Insurer and an insured should not be mistaken as being the same as between litigants, although the claim can produce litigation. Insurers will want to give their clients a full opportunity to challenge any conclusions that result from a forensic engineering investigation before they harden their position. An Insurer will want to know from the forensic engineer what doubts he/she has about his/her. If an Insurer decides to repudiate a claim on the strength of forensic evidence, the last thing they will want is for the repudiation to collapse later when a flaw is found in the evidence. By this time huge legal costs may have mounted and severe damage been done to the relationship between Insurer and insured.

INSURERS' OBLIGATIONS TO THIRD PARTIES IN LIABILITY CASES
Third parties - treated differently from insureds

Although Insurers have obligations to the parties they insure (by virtue of the policy or for commercial reasons) no such obligations exist between an Insurer and a party who is claiming against an insured. This is a distinction that is easy to miss. Third party claimants will generally get nothing unless they can provide hard proof of whatever they say.

Forensic engineers sometimes press ahead with investigations paying little regard to the rights of other parties who are, or will be, engaged in legal proceedings. Some evidence may only be available for a limited period. If a third party is not given an opportunity to examine such evidence and it is then destroyed or rendered inaccessible (e.g. by being buried) they will use this to challenge the validity of any conclusions built upon that evidence. They have been denied their right to have their own expert examination. It is vital therefore for the forensic engineer to be conscious of his/her client's obligations to third parties and to check if others should be invited to inspect evidence at the appropriate time even if this slows down the investigative process.

Striking a balance

For some third party claims there is a clear case for settlement. Others need to be resisted. Between them are the cases where most forensic engineering work is likely to be needed. If it appears unlikely that a case will go to court, Insurers may not wish to volunteer information that might weaken their negotiating position. They therefore face a balance between on the one hand revealing too much and on the other possibly prejudicing the third party's rights. The forensic engineer is often the pivot of this balance. He/she can find himself/herself having to tread a fine line between looking after the Insurer's (his/her client's) interests and compromising his professionalism. This fine line is simply a fact of life. Some forensic engineers find it difficult to identify the line and disagreeable to be made to walk it. If a forensic engineer is not under a duty to the courts to tell the whole truth and does not have any duty to the claimant he/she is at liberty to select which facts to present if asked to in a report - provided this does not result in deception. For example a forensic engineer may be asked to set out the grounds on which the cause of loss could be considered to be negligence.

He/she may omit from his/her report whether he/she believes the loss to have been the result of negligence. In simple terms he/she could limit his/her report to the question:

> Could negligence have been caused this loss and if so how? – Rather than "Do you think this loss was caused by negligence?"

 It will be for the claimant and his/her advisers to probe the forensic information provided. If at a later stage in a dispute a forensic engineer is put under a duty to a court, or to both parties, the nature of a report he/she produces may have to change to include things that were previously omitted. If he/she does not want to submit an "edited" report which might be sent to the claimant he/she can always help draft a letter to be sent by the Insurers to the claimant quoting extracts from his/her findings.

INSURERS AND LAWYERS
Attack, defend or compromise?
Insurers have to work with lawyers a great deal. Particularly if a claim might be repudiated, Insurers may brief a lawyer at an early stage to avoid prejudicing their own position, often to preserve privilege over evidence and to assist in weighing the merits of a case.

Lawyers are generally defenders or attackers of other parties rather than compromisers. Forensic engineers working to the instructions of lawyers need to provide balanced advice on their investigations (Ward 1999). It is not useful for a forensic engineer to give an insurer an over-optimistic interpretation of the evidence. This could lead them into fighting a case that it would be more economic to settle.

Houses of cards
It is easy for a forensic engineer to be drawn into producing a hypothesis as to what happened and to develop it to a sophisticated level based upon an assumption that, when challenged, proves inappropriate. It is therefore the job of the forensic engineer to identify and test any assumptions he/she uses and explain the strengths and weaknesses of those assumptions to his/her client. It may be that only the forensic engineer who will appreciate what assumption he/she has had to make. A forensic engineer's opinion built on an unsafe assumption may be rapidly destroyed under cross examination, damaging both his client's case and his own reputation.

COMMERCIAL CONCERNS
Strange clients?
Many forensic engineers who work with Insurers for the first time find the behaviour of their clients strange. They may find a watertight case is conceded with no logical explanation. What the forensic engineer may not be aware of, is the complex web of commercial interests within which an Insurer is working. Insurance policies are far from straightforward and in addition to the insured often involve many other parties such as:-

Direct Insurer -	the company that issues the primary policy
Local Reinsurer -	a reinsurer that operates in the same country as the insured
International Reinsurers -	a reinsurer operating internationally
Local Brokers -	the broker operating in the same country as the insured who deals directly with the Insured
Reinsurance Brokers -	the broker who arranges reinsurance for a local insurer or local reinsurer

Following Insurers/Reinsurers -	Insurers/Reinsurers who take a proportion of the risk but who are not primarily in control of a claim, although they may influence decisions
Captive Insurers	Insurers owned by the Insured that frequently operate offshore
Treaty Reinsurers	Reinsurers who provide reinsurance cover to direct Insurers through a treaty covering a spread of policies rather than an individual policy

Cases may involve more than one insurance policy and so the number of parties can multiply. Insureds, Insurers, Reinsurers and Brokers frequently work with each other and thus there are bound to be numerous occasions when decisions will make sense for commercial reasons which would not seem sensible when viewed in isolation.

Behind the scenes
Policies are written in layers with different excesses. These complexities mean that it is not at first obvious which Insurers will pay what proportion of a loss. With captive arrangements, for example, an insurer may only be managing the handling of a claim for a subsidiary of the insured. Such a claim may appear initially to be dealt with in the same way as any other claim, but the Insurer may end up paying nothing, and will only decide how much the insured's subsidiary should pay.

INSURANCE COVERAGE ISSUES RELEVANT TO FORENSIC ENGINEERING EVIDENCE
Questions initially asked of forensic engineers frequently are:

- Was the cause one of those that the policy lists as being insured?

- Was .... (an action, a design, a response to a problem) reasonable or not?

- Was the loss caused by negligence?

- Was the disaster foreseen?

- Did defective design, workmanship or materials cause the loss?

But for the forensic engineer to be able to provide a valuable service it is vital that the question he/she is being asked to answer is far more clearly defined than this. It is often valuable to spend some effort defining exactly what an Insurer means by "reasonable" or "negligent". For example evidence of negligence may not be enough to deny a claim because the insurer has to establish that gross negligence was involved. The forensic engineer also needs to know against what criteria he/she is being asked to judge his views on reasonableness. Time invested in tuning the wording of a question almost always pays dividends. Sometimes the tuning can produce a question to which a straight Yes or No is the response and this simplicity should not be scorned.

References

Gloyn B (1999) – "The Unrecognised Importance of Insurance" – Forensic Engineering: a professional approach to investigation, Thomas Telford, London pp. 11-18.

Ward J (1999) – "What are lawyers looking for?" - Forensic Engineering: a professional approach to investigation, Thomas Telford, London pp. 19-26.

Who Pays?: Culpability of experts for building failures in Singapore

DR ANNE MAGDALINE NETTO
Assistant Professor, Department of Building, School of Design & Environment, National University of Singapore, Singapore.

DR ALICE CHRISTUDASON
Associate Professor, Department of Real Estate, School of Design & Environment, National University of Singapore, Singapore.

INTRODUCTION

This paper examines the legislation in Singapore which primarily intends safeguards against latent defects through the implementation of stringent checks with regard to the design and construction process. These checks are required to be carried out by "independent" qualified individuals who are experts from the civil and structural engineering professions. In the light of this, the last section of this paper deals with Singapore's position on the recovery of pure economic loss from building experts in the tort of negligence.

COLLAPSE OF HOTEL NEW WORLD IN SINGAPORE

The legislation was enacted following a major calamity which occurred on 15 March 1986. A six-floor tenement, known popularly as the Hotel New World, collapsed unexpectedly, killing 33 people and injuring many more. More accurately, the building in question was the Lian Yak Building, which had housed the Hotel New World, prompting the former reference to this tenant in media reports even as news of the disaster reverberated throughout the country. In any case, the subsequent inquiry into the catastrophe named their findings as "*Report of the Inquiry into the Collapse of Hotel New World*".[1]

The findings of the inquiry traced the derivative source of the collapse retrospectively to "when the structural design was still on the drawing board" placing responsibility squarely on the expert responsible for structural design. In addition, the following causal factors were also attributed as critical determinants, namely:[2]

- the unsatisfactory quality of the construction;
- substantial loads that were inadequately provided for in the structural design but were added to the building, and caused the overloaded and poorly constructed structure to be further burdened;
- lack of proper maintenance of Lian Yak Building.

---

[1] Mr Justice L.P. Thean, Dr A. Vijiaratnam, Professor S.L. Lee and Professor Bengt B. Broms, "*Report of the Inquiry into the Collapse of Hotel New World,*" February 16, 1987.
[2] Ibid. at pp. 62 to 63.

Having identified that the root cause of the problem had been at the design stage, it was recommended[3] that "all structural plans and calculations of a building prepared by a professional engineer for submission to the Development and Building Control Division ("DBCD")[4] should be checked by another[5] professional engineer, crucially noting that it was to be *an independent check*. The inquiry determined that checks in respect of the main structural elements such as foundations, columns, beams and shear cores were imperative.[6] In order for effective and efficient checks to be carried out, the inquiry advocated that a professional engineer carry out such inspections, who would possess a requisite stipulated number of years' experience in the relevant areas of design and construction of buildings.

LEGISLATIVE AFTERMATH

In response to these findings of the inquiry, the Building Control Act 1989 (Cap 9) was enacted concurrently with the Building Control (Accredited Checkers) Regulations 1989. The fundamental feature of the legislation was the creation of the role of an "accredited checker".[7] The accredited checker serves to provide "an additional"[8] level of "quality control" in the process of design.[9] The legislation exacts on every person for whom building works are to be carried out the prerequisite to appoint an accredited checker. As building works are usually carried out for an employer/developer, the accredited checker would invariably be affiliated to the employer/developer who had appointed him. The accredited checker would also have to be registered with the Building Authority and must further maintain no professional or financial interest (other than the stipulated appointment) in the building works concerned.[10] The legislation also stipulated that only qualified civil or structural engineers of 10-years standing in terms of practical experience in the design and construction of buildings, in addition to being distinguished by ability, standing or special knowledge or experience could be appointed as accredited independent checkers. Evidently, the legislation attempted to ensure that the professional stature of the expert would safeguard his independence when appointed as accredited checker. A system of technical control had thus been put in place with this creature of legislation.

Section 6[11] of the legislation authorises the Building Authority to reject building plans that are not properly prepared and to require the structural designs of any building works to be inspected by an accredited checker. The Building Authority could accordingly, endorse all structural plans of building works on the basis of the certificate and evaluation report of the accredited checker. The Building Authority also has the power to carry out random checks with respect to structural plans and design calculations of the building works. The Building

---

[3] Ibid. at p. 78, para. 12.4.

[4] This was the government department that controlled the planning and development of properties and construction of buildings in Singapore at that time. The department that approves structural plans is now called the Building Engineering Division. The Building Plan and Management Division approves all other plans. Both divisions are part of the Building and Construction Authority ("BCA"). This new statutory board was created on 1 April 1999.

[5] There had already been in place a control regime which was imposed on the 'qualified person' appointed by the building owner. A registered architect or engineer, the qualified person has the responsibility of preparing and obtaining approval of plans from the building authority.

[6] *Report of the Inquiry into the Collapse of Hotel New World* (1987) at p. 78.

[7] Section 6 of the Act.

[8] Additional because the DBCD, persevered in their role as an approving authority on building plans.

[9] Nigel M. Robinson, Anthony P. Lavers, George K.H. Tan and Raymond Chan *Construction Law in Singapore and Malaysia* (1996) Butterworths, at p. 143.

[10] Section 15.

[11] As amended by the Building Control (Amendment) Act 1995, s. 5.

Authority thus retains the right to revoke acquiescence of the building plans if satisfied that any information given in respect of the approval had been false in a material particular.

Section 33[12] insulates the government, the Building Authority and any public officer from suit arising from information provided to the public by the Building Authority by electronic or other means.[13] Accordingly, the Building Authority has been given unequivocal protection which the landmark decision of *Murphy v. Brentwood District Council*[14] achieved in England in 1991.[15]

EFFICACY OF LEGISLATIVE REQUIREMENTS

This legislative mechanism introduced in Singapore is not without problems. Although the intention of the revised legislation had been to provide additional independent checks for buildings works, it is questionable whether the accredited checker is truly independent, as it is the employer who is expected to engage him. As a professional engineer accredited checker, he would undoubtedly be exposed to liability if he fails to discharge his obligations to the standards expected of him. In that regard, the legislation demands the accredited checker to state in the certificate issued by him that "to the best of his knowledge and belief, the plans so checked do not show any inadequacy in the key structural elements of the building to be erected or affected by building works carried out in accordance with those plans."[16] In addition, the accredited checker faces criminal sanctions under the legislation that created him.[17] It should also be noted that the expert engineer could be liable in the tort[18] of negligence for failing to discharge his duty of care to the employer/developer and even to subsequent owners of the building.

Following the enactment of such stringent provisions, many Singaporeans lent the view that no further protection from delinquent planning or shoddy workmanship, especially in relation to structural design would be required for building and property owners. The standard, almost premonitory retort that has been offered, has been that the legislative system could not guarantee that there will be no further accidents, this system's success being dependent on the diligence of the experts placed to control and safeguard building standards.

As if in legislative rejoinder, in June 1999, the roof of an unfinished building[19] collapsed. Fortunately the building, which was to be a school's multi-purpose hall had been uncompleted. Seven workers were killed in this disaster and the multi-purpose hall was left in ruins. The builder, engineer and accredited checker all pleaded guilty to criminal charges under the Building Control legislation. The accredited checker admitted that he had been remiss in inspecting the engineers' structural plans and the design calculations of the buildings and that he had also failed to do an independent calculation to determine the

---

[12] Ibid. s. 21.

[13] See *Swee Hong Investment Pte. Ltd. v. Swee Hong Exim Pte. Lt.d & Anor* [1994] 3 SLR 320. See also "Building Control (Amendment) Act 1995" (1995) Singapore Academy of Law Journal 455.

[14] [1991] 1 AC 398.

[15] See *Dutton v. Bognor Regis U.D.C.* [1972] 1 QB 373; *Anns v. Merton London Borough Council* [1978] AC 728 and *Governors of the Peabody Donation Fund v. Sir Lindsay Parkinson & Co. Ltd.* [1985] AC 210. For a comparative study review see also, A.P. Lavers and N.M. Robinson, "Building Control: A review of the Singapore Approach" (1990) 7 ICLR 111.

[16] See section 6(1) (c).

[17] Section 18 which provides for a fine and jail sentence.

[18] *RSP Architects Planners & Engineers (Raglan Squire & Partners) F.R. v. Management Corporation Strata Title Plan No. 1075* [1999] 2 SLR 449.

[19] The Compassvale Primary School and referred to hereafter as 'Compassvale'.

structural adequacy of the multi-purpose hall. On 20 March 2000, the district judge, Tan Puay Boon dealt the maximum fines to both the engineer and accredited checker. Engineer Bill Hong was fined a total $25,000 for flouting building control laws and accredited checker, Joseph Huang was fined $50,000 in the light of the demonstrable incompetence in verifying structural plans and design calculations for the building work. The building contractor, BKB Engineering Constructions was fined $30,000 for failing to ensure that the thickness of the weld at the roof joints which held up the roof complied in actuality with the requirements in the approved structural plans.

It is therefore clear that the technical control regime is manifestly inadequate in its laudable aim of preventing the occurrence of avoidable accidents largely due to egregious human negligence or professional misconduct, or for that matter, in discernibly eradicating the existence of latent defects in buildings. The extent to which the building control legislation has been blatantly flouted is evident in the case of Compassvale; it is noteworthy that the collapse of the roof took place *even before* the building saw its *completion*. In that sense, the Compassvale disaster is culpably more serious than the Hotel New World disaster and exposes serious shortcomings and oversights in the design and building procedures. It may be argued convincingly however, that only one failure since the implementation of the Act is patently sufficient evidence of the success of this particular legislative mechanism. Indeed, it needs to be conceded that there has to be some margin for human error. The allowance for this degree of error does, however, make a further case for a back-up insurance scheme, such as the decennial system, which boasts of a truly independent technical control system.

CIVIL LIABILITY OF EXPERTS IN TORT

In 1995 the Singapore Court of Appeal had an opportunity to examine the tort of negligence in the context of latent defects in buildings. In *RSP Architects Planners & Engineers v. Ocean Front Pte. Ltd. & Anor*[20] (referred to hereafter as "*Bayshore Park*"[21]) the Singapore Court of Appeal had to decide on the issue of pure economic loss arising from defective buildings.

The Plaintiffs in this case were the Management Corporation of Bayshore Park condominium. The Defendants were Ocean Front Pte Ltd., the developers of that condominium. The Defendant developers joined RSP Architects Planners & Engineers as third parties.

The developer was being sued for faulty construction of the common property which had led to spalling of concrete in the ceilings of the car parks of the various blocks and water ponding in the area surrounding the lifts. One of the issues which arose was whether the management corporation was barred from claiming for pure economic loss. Thean JA held *inter alia* that there was a duty on the part of the developers to take care to avoid causing to the management corporation the kind of damage the latter had sustained and that there were no policy reasons to negative such a duty of care.

In coming to the above decision, the Court chose to adopt the *Anns v. London Borough of Merton*[22] two-stage test as applied in *Junior Books v. Veitchi*[23] and refused to follow *Murphy v. Brentwood District Council*.[24]. Thus, recovery was allowed for pure economic loss[25].

---

[20] [1996] 1 SLR 113.
[21] As this is the name of the private residential development in the case.
[22] [1978] AC 728.
[23] [1983] 1 AC 520.
[24] [1991] 1 AC 398.

The ***Bayshore Park*** decision was recently applied in ***Management Corporation Strata Title Plan No. 1075 v RSP Architects Planners & Engineers (Raglan Squire & Partners) F.R***[26] (hereafter referred to as ***"Eastern Lagoon II***[27]*"*). In this case, the Plaintiffs were the Management Corporation of the condominium known as Eastern Lagoon II. The defendants were an architectural and engineering firm. The main contractor for the condominium was brought in by the defendants as a third party.

A panel of bricks and brick tiles at the fifth storey of one of the tower blocks in the condominium fell onto the roof of a unit in another block. The plaintiffs claimed that the defendants had been negligent in designing and supervising the construction of the condominium. The defendants brought a third party action against the main contractor for the condominium seeking a contribution/indemnity from them. The third party's response was to assert that the defects arose from the defendant's default and not their own.

The issue which arose was whether there was a duty of care on the part of the architects to exercise reasonable care in the design and supervision of the construction of the common property so as to guard against the plaintiffs sustaining the damage complained of ie., rectification of defects caused by negligent design and/or supervision.

Judith Prakash J. in the High Court answered this in the affirmative for several reasons, including the fact that it was "obviously foreseeable by the defendants that if they were negligent in the design of the condominium, this could result in expensive rectification work and therefore economic loss for either or both the subsidiary proprietors and the MC". She used the reasoning in Bayshore Park to hold that there were no policy reasons to negative the duty of care and found for the plaintiffs by allowing recovery for economic loss. It is clear that the learned judge was bound by the earlier Court of Appeal decision of ***Bayshore Park.***

On appeal,[28] the Singapore Court of Appeal was given an opportunity to re-examine its position on the recovery of pure economic losses in the context of latent defects in property. Counsel for the defendants were faced with having to overcome the previous Singapore Court of Appeal decision in ***Bayshore Park***, which had clearly allowed for the recovery of pure economic losses. On the facts, the defendants' case had no merit. Their counsel faced an uphill task as ultimately, there had to be overwhelming reasons *not* to find the negligent defendants liable. This was especially difficult as the *safety* of persons and property was in issue. It was therefore not surprising that the Court of Appeal in ***Eastern Lagoon II*** decided in favour of the management corporation and allowed recovery for pure economic losses in line with ***Bayshore Pa***rk.

CONCLUSION
The above discussion has shown that the technical control regime which has been in put in place[29] is manifestly inadequate to prevent the occurrence of building defects arising from the negligence of professionals involved in the construction process. The Compassvale

---

[25] See A.M. Netto and Dr Alice Christudason, "*Junior Books* Extended" (1999) 15 Const LJ 199 for a discussion of cases on the point.

[26] Suit No. 1260 of 1995, judgment delivered on 9 September 1998, unreported.

[27] As this is the name of the private residential development in the case.

[28] *RSP Architects Planners & Engineers (Raglan Squire & Partners) F.R. v. Management Corporation Strata Title Plan No. 1075* [1999] 2 SLR 449.

[29] The Building Control Act 1989 (Cap 9) and the Building Control (Accredited Checkers) Regulations 1989.

collapse illustrates that there are serious shortcomings and oversights in the statutory controls impsed on design and building procedures. Although this is the only case which has exposed the legislation's inadequacies, it is still one case too many.

However, despite the legislative scheme, at present Singapore, as with the other Commonwealth common law jurisdictions and unlike England, allows for the recovery for pure economic losses for buildings. This may offer some consolation to an aggrieved owner of a defective building .

A selection of forensic engineering cases

DAPHNE WASSERMANN
Cadogan Consultants, Glasgow, UK

INTRODUCTION

All sorts of enquiries pass across the forensic engineer's desk. In this paper I will give a taste of the variety of subject matter and approach needed in this work. More detailed consideration will be given to the failure of a bellows unit on a blast furnace compressed air line. The case involved several experts and was interesting in that the obvious factor was not the correct answer. Brief accounts of other cases follow.

BELLOWS FAILURE

Background

Figure 1 is a schematic of the system. Air from either blower 1 or blower 2 is compressed, passing along a 1600mm cold blast main at 300°C and about 6 Bar. The compressed air is heated in one or more of four stoves. Hot and cold air are mixed to the correct temperature before passing to the blast furnace.

The blast furnace contains iron ore and coke, charged at the top. Within the furnace oxygen in the air reacts with coke to form carbon monoxide and steam forms carbon monoxide and hydrogen. These gases reduce the iron ore to molten iron. Impurities form a molten slag layer above the iron. Gases pass up the furnace and the blast furnace gas, a mixture of hydrogen and carbon monoxide exits at the top. Molten iron and slag are bled off from time to time.

The pressure of gas entering the blast furnace is essential to keep the charge in place and allow gas flow to through it.

Bellows units are situated along the airline to provide flexibility as the line heats and cools. The bellows unit consists of a flexible convoluted section. Internal pressure in the pipeline generates an axial load along the pipe. To counteract this load, the flanges at the two ends of the bellows unit are connected by a hinge arrangement (see figure 2) allowing rotation in one plane.

In the present case, the welds between the hinge and the flange failed causing failure of the bellows unit. The air pressure was lost and as a result the burden in the blast furnace fell into the pool of molten metal and slag. These were forced back down the airline where they solidified. The consequential losses were estimated at several million pounds. At the bellows end, whipping of the airline into the adjacent control room caused minor injury of an operator.

Forensic engineering: the investigation of failures. Thomas Telford, London, 2001.

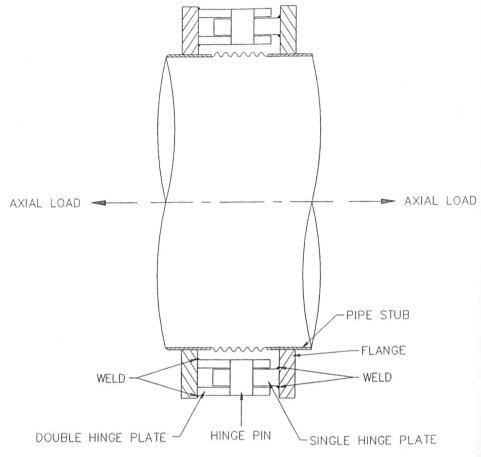

FIGURE 2.   CROSS–SECTION THROUGH BELLOWS UNIT

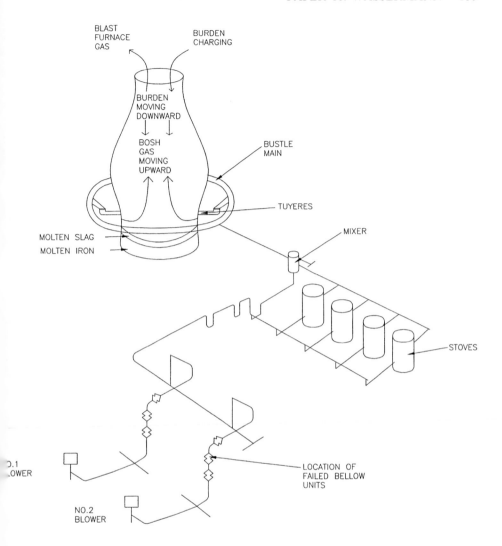

FIGURE 1. LAYOUT OF BLAST MAIN AND BLAST FURNACE

Cause of Failure: Plaintiffs' Case

The Plaintiff as operator of the blast furnace, took out an action against the bellows manufacturer. The Plaintiff's metallurgist examined the failed welds and noted that the root gap was larger than it should have been. They concluded that the weld was poor and that this must have caused the failure. They did not initially examine other operational aspects at the time of failure, nor investigate the previous operating history of the plant.

Retaining the evidence

The Plaintiff carried out metallurgical examination of the weld. They cut sections through the failed weld but unfortunately did not retain the full weld. Some computer records from the time of the failure were also not retained. This lack of evidence hampered the investigation and did not help their case.

The Defendant's experts

The bellows manufacturer, as Defendant, retained experts in three areas: blast furnace operation, fracture mechanics and ourselves as mechanical engineers. Between the experts a much fuller coverage was able to be given of possible causes of failure. The Plaintiff also employed several experts.

Failure mechanism

The bellows manufacturer had been given a temperature and pressure for the design of the units. He had not been informed of any dynamic loads, nor of a corrosive environment. The fracture mechanics expert examined the retained sections of the weld. Although the root gap was large, this had been compensated for by additional weld material. Static analysis showed that 75% of the weld length would have to be removed before the weld would fail under the static load. Indeed, examination of the operational history confirmed that the unit had withstood overpressurisation without failing.

Examination of the metallurgical specimens that were retained from the unfailed ends of the hinges showed that some areas of weld with small cross-section had not failed while some areas of large cross-section were cracked.

The fracture mechanics expert considered the following evidence
- The cracks were tortuous in texture
- The crack surfaces were heavily corroded
- The bellows unit operated for over 10 years before failure
- The plant was subject to pressure fluctuations
- The bellows unit operated at 200°C.

This led him to suspect fatigue failure as the mechanism. Detailed investigation of the operating history was undertaken by examining records of the pressure in the two lines. This revealed that blowers were changed over every 16 days, giving perhaps 25 major cycles per year. In addition there were numerous smaller and larger fluctuations about a high mean pressure.

An additional factor was the proximity of the installation to the sea, which was about 1km away in two directions. The installation would therefore be in contact with salt-bearing air.

The expert carried out accelerated fatigue tests on samples of similar material and demonstrated that with the cycles alone the weld life would have exceeded 25 years.

Corrosion alone would also not have caused failure. However, fatigue failure is accelerated in the presence of a corrosive environment such as hot salty air. This combination led to accelerated failure by corrosion fatigue.

In the absence of the full weld, the location of the initiation site and the path of crack propagation could not be determined.

Maintenance
There are two hinges on the bellows unit both of which had to fail to cause the bellows unit to fail. The levels of corrosion suggested that one of the welds had failed completely at some considerable time before the accident. The second weld had taken the complete axial load for a number of months without failing.

Although the bellows unit was several metres above ground level it could be observed visually and could have been inspected with binoculars. In addition, cracking of the first weld which failed was almost certainly a gradual process. No maintenance inspections of the bellows unit had been carried out in the ten years of its operation. The compressed air line itself had only had occasional inspections, although it was known to be a critical item.

Operational Difficulties
There had been problems with the parts of the line near the blowers in the first few years of operation. Measurements revealed excessive noise levels which caused unacceptable vibration and failure of the adjacent silencers. The bellows units had been modified by removing the inner sleeves to prevent damage from the noise-induced vibration. The Plaintiff was therefore fully aware of unanticipated loads on this part of the compressed air line.

Events immediately preceding the failure
The blast furnace expert examined the operational records on the day of the failure. In his opinion, there was evidence of operating difficulties resulting in a loss of gas flow up the blast furnace. The gases may have reversed their flow, entering the airline and causing an explosion a few minutes before the bellows failure. The pressure surge from the explosion could have placed additional loads on the bellows unit causing it to fracture.

Charts of pressure and temperature supported this view. Computer records of gas composition had been deleted.

Summary
The Plaintiff made assumptions without a full scientific evaluation. They subsequently threw away most of the failed component.

Their logic was flawed. They assumed that if the weld was poor then it was the cause of the failure. Yet in the unfailed welds areas of smaller weld cross-section were intact while larger cross-sections were cracked.

The Plaintiff had failed to take into account the salty environment and the probable vibrational loads and had not specified these to the bellows manufacturer.

The Plaintiff also failed to look at the operational conditions at the time of failure to see if anything abnormal could have assisted the failure.

The initial design with a single air supply was known to carry risks. Yet maintenance was poor so that the incipient failure was not detected.

By looking beyond the obvious we were able to find evidence of a variety of other factors that led to different conclusions about the failure mechanism and the liability for the failure.

The Plaintiff lost the case.

CONTRIBUTORY NEGLIGENCE
Many cases involve weld failures. An example from a different location was an accident involving a small lorry - mounted crane. This is the type of crane where the operator sits on a seat which forms part of the crane upright. The crane is used to move goods on and off the lorry or from one part of the lorry to another. In this case the weld at the base of the crane failed, the crane and operator fell off the lorry and unfortunately the operator sustained injuries to his back. Again, the weld was poor, but calculations suggested that nevertheless, it should not have failed if the crane had been operated correctly. Photographs of the lorry taken just after the accident suggested that the outriggers were not down. If the lorry was not stabilised by the outriggers it would tip as the load moved and this would increase the stresses in the welded joint. Nevertheless, the Plaintiff was successful.

PERSONAL INJURY
Most cases of personal injury involve a single expert and a relatively small claim. While engineering or technical elements are usually present there may be other considerations such as health and safety requirements and safe systems of work.

An employee at a glass bottle making factory suffered an injury to his hand when it was caught in an glass bottle making machine which started to operate unexpectedly. The employee raised a civil action against the employer.

An engineering consideration of the control system demonstrated that four separate control and safety devices would need to malfunction to allow the machine to start unexpectedly. On a balance of probabilities this was extremely unlikely. The judge was of the opinion that the injured party must have contributed to his misfortune.

A further lesson was learned from this case. The valves could not be dismantled for inspection and the engineer assumed that they were normal straight-through valves on which the handle turned through 180° from closed to open. In court, the valve was presented to the engineer. Inspection showed that it was of an unusual type with a 90° bend in the port, requiring a turn through 270°C to open it. It is dangerous to make assumptions without checking the facts.

PHOTOGRAPHIC EVIDENCE
An example of personal injury where photographic evidence can be misleading involved a man who fell from scaffolding. It transpired that the scaffolding was some distance from the surface on which he was working. While reaching out he fell and was killed. During the course of the investigation photographs were taken which demonstrated the dangers of relying on photographic evidence. From one angle the photograph suggested that the scaffolding extended beyond the building. Photographed from another angle it was clear that it did not. The case settled in favour of the deceased's widow.

The colour on photographs can be particularly misleading. On one photograph a weld surface may appear to be corroded and rust coloured. Taken with a different film or lighting the surface may appear grey and clean.

When taking photographs that are to be used as more than just indicative, care should be taken to note all the details such as camera settings, distance from the object, lighting, weather conditions. Where appropriate a rule or other object to indicate size should be included in the view.

If photographic evidence is critical it is preferable that the photographs be taken by a professional forensic photographer.

POLLUTION INCIDENTS

Engineering opinion may also be sought in cases where the damage has been to the environment or to animals rather than to people or equipment

Oil Spillage

One case involved oil leakage into a stream. A demolition company removed half the bund from a disused heavy fuel oil tank before thinking to check whether the tank was empty. On discovering a residue of oil in the tank they called in specialist tank cleaners. The latter chose to clean the tank by dissolving the heavy fuel oil in kerosene. About two days after the kerosene had been added, pollution of a nearby stream was noticed and the Regulator was alerted.

It transpired that there was a cut in a pipe which ran from the tank at low level. Once the kerosene had dissolved the tarry heavy fuel oil blocking the pipe the kerosene / oil mixture had poured out of the cut. The Regulator was of the opinion that the oil leaked out from where the bund wall had been removed and that the fault lay with the demolition contractor. However, a video survey of the drains on the site suggested that the drains on that side of the bund were clean and blocked with debris. The oil path appeared to be from the other side of the bund wall which was clearly not water tight. Even if the bund wall had not been demolished the leak would probably not have been prevented.

Further calculations using the viscosities of heavy fuel oil and kerosene showed that without the introduction of the kerosene no leak would have occurred. Given that the tank was old and some pipes had already been removed, the tank cleaners could not have expected that the tank would necessarily be intact. They must therefore bear considerable responsibility for the spillage.

Air Pollution

At the time, this trial was the longest running civil case in the UK. A farmer's land was situated adjacent to a special waste incinerator. Some of his cows died and he claimed that they were poisoned by pollution from the incinerator.

Extensive analysis of charts of smoke emissions from the incinerator was undertaken. Operational records also showed that on some occasions the wet scrubber was not working and un-cleaned gases were being discharged. Temperature records also showed evidence of poor operation and witness statements confirmed this. We were therefore able to demonstrate that significant pollution, particularly dioxins, could have left the incinerator.

Other experts carried out plume dispersion modelling and looked at meteorological data. They showed that the dioxins could have reached the farmer's field. Veterinary experts gave opinion on whether the cattle died from dioxin poisoning. The case is an example of a chain of events, all of which need to be proven. If one link is broken the chain itself is no longer intact and in the event the Plaintiff's case failed.

PLANT AND PROCESS PERFORMANCE
Not all disputes arise from catastrophic failure. A plant may not perform in the way that the purchaser intended. The specification may require a certain input of raw materials and energy for a given output or quality of product. Even if there is no process failure, litigation can result if the guarantees are not met. Examples include a fertiliser plant and a polyester factory. The problems are exacerbated if new plant is retro-fitted onto older parts or if the supplier of new equipment is not in control of all the inputs.

CONTRACTUAL DISPUTES AND SPECIFICATIONS
The above example demonstrates the importance of clearly written specifications. In some cases there may be no specification at all. In other cases it may be poorly written or contradictory.

In a recent case a contractor was to design and install a combined heat and power (CHP) system in an existing building, to integrate with existing CHP, boilers and power supplies. Although the Contractor had denied responsibility, the contract, using MF/1, was written in a way which would suggest that the Purchaser had done much of the design. However, critical areas had been omitted, such as full discussions with the Regional Electricity Company regarding export and import. Outline plant layouts were supplied which would not work in practice. The standard appendices gave detailed requirements for building services ducting but no information on high temperature exhaust gas ducting. The specification stated that the detailed design should comply in all respects with the Purchaser's requirements and also stated that the specification covered all works involved. This was not the case and led to ambiguities about the scope of supply and the responsibility of the parties.

PLANNING INQUIRIES
In all the cases described so far the Engineer is appointed after something has gone wrong. A public inquiry into a planning application has a different purpose. The presumption is in favour of the developer. The objector, often the local authority spurred on by protest groups, must think of all the things that could go wrong and what the harmful effects might be. The developer must anticipate possible objections and show how he proposes to mitigate harm. Most of our inquiries have concerned incinerators, a very emotive subject. In these cases practical experience as an engineer of all the difficulties of operating the plant in question is essential.

CONCLUSION
The variety of enquiries is enormous. In approaching them it is important to keep an open mind. The Engineer must go beyond the obvious solution to consider competing probabilities. Imagination in identifying and exploring all possibilities is key.

Post-Mortem Seismic Evaluation of an Office Building

ROBIN SHEPHERD, Ph.D., D.Sc., F.I.C.E., F.ASCE.; President,
Earthquake Damage Analysis Corporation, Big Bear Lake, California, U.S.A.

ABSTRACT

This paper describes the comprehensive autopsy conducted on a two story office building that was severely damaged in the 1994 Northridge earthquake. The investigation was undertaken as part of an earthquake insurance claim and involved extensive destructive and non-destructive testing.

Estimates of the accelerations experienced by the subject building indicated that significant structural damage would be expected in this apparently code compliant structure. The detailed observations of the behavior were in accord with such expectations. The ability of the building to resist future earthquakes was substantially diminished by the damage suffered in the Northridge earthquake. The insurance claim, resolved by binding arbitration, resulted in $M 1.1 being awarded but the insurance company declared bankruptcy before payment was made.

THE EARTHQUAKE

Although only a moderate event in terms of energy release, the Magnitude 6.7 Northridge earthquake that occurred at 4:31 a.m. on January 17, 1994 was noteworthy as it occurred directly below a large metropolitan area and caused very extensive damage to constructed facilities (1 & 2). The duration of the strong ground motion was between fifteen and twenty seconds and peak ground accelerations having horizontal components as high as 0.9g were measured in several locations with vertical accelerations of the order of two-thirds the corresponding horizontal ones. Many complete or partial structural collapses occurred, including wooden apartment buildings, older cast-in-place concrete buildings, precast concrete parking buildings, and reinforced concrete freeway structures. In the circumstances of the severe ground motions recorded over a significant part of the San Fernando Valley, the fact that there were only some sixty casualties resulting from the structural damage was consistent with the intent of the seismic design code, namely life protection.

The subject building was approximately seven miles southwest of the epicenter and the site is estimated to have experienced an effective peak (horizontal) ground acceleration of 0.35 to 0.4 g with accelerations in the superstructure of around 0.7 to 0.8 g horizontal, with vertical ones of corresponding order of magnitude.

Forensic engineering: the investigation of failures. Thomas Telford, London, 2001.

THE BUILDING

The building was a two story structure (Figure 1), with a footprint of approximately 474 square meters on the lower level (Figure 2) and 538 square meters on the upper one.

Figure 1.   View of building from the south west.

At the time of the Northridge earthquake it was about ten years old. The construction comprised concrete perimeter foundation walls with a slab-on-grade lower floor, timber framed walls and built-up timber floor system on the upper level having plywood, nailed and glued to the joists, covered with a lightweight concrete slab.

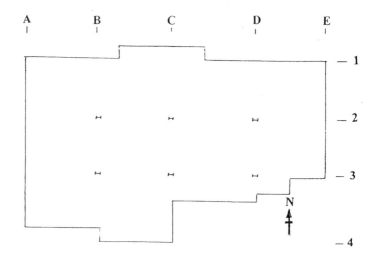

Figure 2.   Footprint of building at ground level.

Steel frames, moment resistant, braced and unbraced, were incorporated into the structural system. The timber frame was stucco clad, with gypsum drywall interior wall and ceiling surfaces. Over the center section of the building the roof was essentially flat plywood with a hot mopped malthoid membrane but the roof had a fringe of red tile on the perimeter built-up sloping section. Small balconies were provided as an architectural feature on the east, south and west faces of the building. The north face was offset near the center of the building.

The lateral-load resisting assemblage of the building was a Dual System per Table 23-1 of the Uniform Building Code, 1982 (3). The primary lateral load resisting elements comprised, at the lower level, steel braced frames (Figures 3 and 4) at each of the east and west faces of the building (on lines A and E, Figure 2). A combination of plywood shear walls on the north and south faces and an internally placed steel moment resting frame (on line 3, Figure 2) provided east-west resistance. At the upper level the primary lateral load resisting elements comprised perimeter plywood shear panels and an internal moment resisting steel frame, on line 3 near to and parallel to the south side of the building.

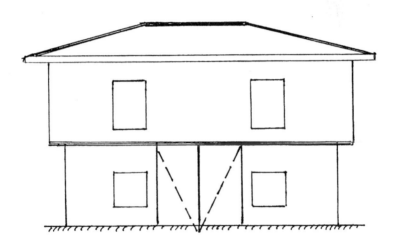

Figure 3. East elevation of building

EARTHQUAKE DAMAGE
Immediately after the earthquake external damage in the form of extensive stucco cracking and internal damage, manifested by drywall cracking, was evident. However, it was not until a more in-depth examination was undertaken that the seriousness of the structural deterioration was established.

A series of inspections, in many of which destructive testing was undertaken, were completed over a period of several months. Damage recorded in addition to cracking on all stucco and drywall faces of the structure, included sill plate movement, slab-on-grade cracking, settlement of internal columns, sloping of the floors, splits (Figure 5) in the upper story nailers (provided to connect the timber upper floor structural system to the top flange of the supporting beams), fractured shear clip connections and permanent displacement of the upper story relative to the lower one.

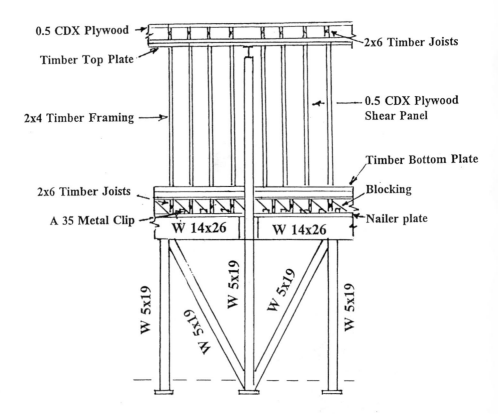

Figure 4. Lateral load path structural components, east and west walls.

In addition to the stucco and concrete cracking evident on all exterior faces of the building, the corbels on the east side were observed to be tilted and twisted. One of the many vertical cracks in the edges of the perimeter of the concrete floor was found to extend more than 50 mm from the surface, well beyond the stucco. Indications of sill plate movement were observed from the outside of the building and such displacement was subsequently confirmed by removal of internal surfaces to reveal cracking in the timber sill plate. The soil was removed alongside the outside wall in several places to determine the depth of the foundation below grade. It was established that the footings were at least as deep as called for on the construction drawings and at the east and west ends of the structure the footings appeared to be deeper and wider than called for on the drawings.

On pulling the floor coverings on the interior of the building it was revealed that the slab-on-grade cracking was severe. The slab was cut and jackhammered to confirm that it did contain both wire mesh reinforcement and bar ties from the foundation to the floor. Removal of drywall revealed splitting of the bottom sill plates in the north wall adjacent to the entrance doorway, in the stairwell, and in the north-south wall east of the stairwell.

Figure 5.   Fractured nailer plate.

Removal of the drywall on the interior of the east wall provided access to the bottom and top of the chevron brace and other steel beam-column joints were similarly accessed by drywall removal. On the basis of visual examination it was considered possible that some of the connections might have experienced post-elastic straining with possible deterioration of the welds. However non-destructive testing undertaken subsequently indicated that no weld damage was found.

The slope of the floor was examined, initially by way of a manometer survey. Local gradients were measured using an engineer's level and two more formal surveys were undertaken, the first to establish the concrete floor elevations around column B-3 (Figure 2) and the second to determine the beam elevations at all interior columns. A differential elevation of more than 25 mm over a one meter horizontal distance consistent with column B-3 footing having moved vertically downwards relative to the slab was measured. The separation of the timber plate from the steel beam adjacent to the top of that column supplied further support to the contention that the column had settled. The later, more comprehensive, surveys revealed that the floor was sloping down towards the column B3 with vertical differences of 16 mm to 20 mm. The survey of the beam elevations at the top of the interior columns confirmed that several of the interior columns had settled.

CORRELATION OF DAMAGE WITH SEISMIC EXCITATION
The overall intent of the Uniform Building Code seismic provisions is to ensure life safety by preventing the collapse of a building when subject to severe ground shaking. This can be interpreted simplistically as providing resistance to a 0.4g horizontal acceleration with significant structural damage but without catastrophic structural failure.

Reasonable estimates of the accelerations experienced by the subject building indicated that non trivial structural damage would be expected in the apparently code compliant building. The observations made of the damage sustained were in accord with such expectations. Specifically the fracturing of the various structural elements such as the nailer plates, sill plates and metal clips, coupled with the permanent sideways offset of the upper level of the building, were consistent with severe earthquake induced lateral loading. Likewise, the damage to the underside surfaces of the upper floor joists where they were bearing on the east wall of the building was consistent with vertical overload.

In response to north-south induced motion the structure would have attempted to resist the inertia forces through the steel chevron braces, at the east and west ends and, until the drywall fractured, through the hallway walls. Additionally, the brick walls that were orientated north-south adjacent to the front doorway, being relatively stiff, would have attracted to themselves in-plane racking loads until they suffered mortar joint failure. As the internal hallway walls had no individual strip footings, being founded on the slab-on-grade, the wall on the eastern side of the hallway would exhibit greater in-plane stiffness as a result of being confined by column bases at either end, namely at positions C2 and C3. This would have resulted in vertical loads being applied to the column bases consistent with the in-plane wall loads.

The effect on the structure of east-west earthquake induced motion would have been to apply inertia forces primarily to the perimeter plywood shear panels on the north face of the building and to the moment resisting steel frame on Line 3. The failure of the sill plate along the north wall is consistent with sequence of events in which this wall became overloaded, and hence severely damaged. The east-west orientated shear wall on the north face, west of the entrance hall, together with the moment resisting frame on Line 3 would then have been called upon to provide a major portion of the seismic resistance in this direction, with consequent incased axial load being imposed on the columns and the ends of the shear walls. The combination of the earthquake induced axial loads generated by both the north-south and east-west movements of the building accounted for the settlement of the C3 column.

The permanent vertical settlement of column B3 was consistent with the attempt of its footing to resist the axial, horizontal and rotational loads induced into that column as a result of east-west motion of the building. The settlement of the north wall west of the hallway was consistent with the increased horizontal force and vertical end-of-wall forces caused by the earthquake when the north wall sill plates failed.

CONCLUSIONS

As a result of the investigation described, the seriousness of the earthquake damage sustained by the subject building was confirmed. The ability of the building to resist earthquakes had been diminished substantially. The load paths necessary to provide seismic resistance had been seriously compromised to the extent that very extensive repairs would have been necessary to restore it to its pre-Northridge condition. It was expected that as more of the building's structure would have been exposed in a repair process, additional damage would have been revealed.

It was anticipated that it would have been necessary to strip the timber floor structure of the upper story to establish the full extent of the damage to the connections. Repair of the damage to the shear connections, and restoration of the deficient joists, nailers, plates and blocking (Figure 6) would likewise have required the upper floor timber structure to the stripped to provide access.

Figure 6. Split block

The repair of the slab-on-grade could have been tackled in two ways. One would have involved removal of the old slab whilst leaving the poured-over-the-top sections in place over the perimeter footing. The second would have involved removing all of the old slab and pouring an entirely new one, extending over the complete ground level footprint of the building. In both approaches it would have been necessary to locate and evaluate damage to the perimeter foundation to determine if repair or replacement would have been necessary.

It was planned to repair foundation wall cracks of appropriate width by epoxy injection except where the width of any crack was greater than that which could be repaired reliably using epoxy, then those areas of wall would have had to be broken out and repoured with dowel connections to the remaining footing concrete.

If the poured-over-the-top sections were to have been left in place it would have been necessary to level the footings at the low spots near the center of the north and south perimeter walls. To ensure a contiguous structural connection between the new slab and the perimeter concrete footing, it would have been necessary to instal rebar dowels to transfer loads both from the footing and from the residual portion of the original slab. The installation of rebar dowels to ensure load transfer would have involved much expensive drilling and bonding. Since the shattered bottom sill plates would have had to be replaced, it appeared more straightforward to strip the building to the steel frame and to pour an entire new slab. Such an approach would have facilitated complete examination of the structure together with the repair of the timber framing and the necessary levelling of the support walls.

INVESTIGATION OUTCOME

The fact that subject building had suffered severe structural damage was established. In particular the ability of the structure to resist earthquakes had been substantially diminished and the risk of collapse in another seismic event was judged to be high.

The vulnerability of weak links in the lateral force resisting load path was clearly demonstrated in the response of this building to Northridge earthquake generated ground motion. The necessity of restoring the connection of the upper level structure to the lower level portion, specifically the replacing of nailer plates, clips, blocking and joists (Figure 4) was the primary factor in resolving the insurance damage claim.

A very real possibility existed that not all of the damage actually sustained in the Northridge earthquake had been discovered in the investigation described. Nevertheless sufficient damage had been identified to cause the mediation panel to conclude that it would be necessary to dismantle the frame of the building from the ground up, to and including the roof, and to start the repair work after such demolition.

REFERENCES

(1)   Moehle, Jack P., (Ed.), "Preliminary Report of the Seismological and Engineering Aspects of the January 17, 1994 Northridge Earthquake, Earthquake Engineering Research Center, University of California, Berkeley.

(2)   Hall, John F., (1995), "Northridge Earthquake of January 17, 1994, Reconnaissance Report", Earthquake Spectra, 11 , Supplement C.

(3)   UBC, (1982), " Uniform Building Code", International Conference of Building Officials, Whittier, California , U.S.A.

Imminent collapse of wood structures affected by decay

Dr. Kenneth B. Simons, Ph.D., P.E., Principal Engineer, Damage Consultants, Inc., Mercer Island, Washington, USA

ABSTRACT

Imminent collapse due to wood decay has become an issue in Washington State, USA. Insurance companies are spending huge sums of money to repair existing wood framed buildings (both multi-family condominium and apartment buildings as well as low-rise commercial buildings and single-family residences) due to case law on the subject of imminent collapse. Typically collapses occur when the structure can no longer support the imposed loads. The cause of these collapses is normally easy to identify due to the combination of the weakened structure (or element of the structure) with a significant event exceeding the capacity of the structure such as a natural hazard (snow, wind, earthquake, etc.) or man-made hazard (improper use, construction activity, alteration). However, collapses often occur when the structure is not loaded with live loads. These types of spontaneous collapses result in extensive studies in order to attempt to determine the cause of the collapse. Most structures (if protected and maintained) including wood frame buildings, can last indefinitely. However, within the last 20 years the combination of modern architectural design, energy conservation methods, new building materials and new wall cladding systems have reduced redundant water resistant features previously present that provided protection for the structure. Thus, wood decay (rotting) of a wood structure, which previously may have taken many years or even centuries in some cases to cause a building to collapse, currently initiates shortly after construction and manifests itself to the extent that portions of a building may be significantly reduced in strength within a few years.

DEFINITIONS

Collapse as defined by Webster's [1986] is "(a) falling away or breaking down; any sudden or complete breakdown or frustration." As a verb, collapse is "to fall or break down; to go to ruin." Imminent as an adjective is defined as "near at hand: impending: threatening." Thus, most would interpret a collapse as an event that has already taken place and that the adjective "imminent" indicates that collapse is near at hand. A comprehensive definition of failure as used by the Technical Council on Forensic Engineering of the American Society of Civil Engineers, Carper [1989] is "an unacceptable difference between expected and observed performance." Carper [1989] also sites Gordon [1978] "In one sense a structure is a device that exists in order to delay some event which is energetically favored. It is energetically advantageous, for instance, for a weight to fall to the ground, for strain energy to be released, and so on. Sooner or later the weight will fall to the ground and strain energy will be released; but it is the business of a structure to delay such events for a season, for a lifetime, or for 1,000 years." Carper [1989] also states that the definition of failure is broad enough to include serviceability problems, and in the case of buildings, the most costly reoccurring performance problems are those associated with building envelope performance. Although leaking roofs and facades are not catastrophic, news making events, forensic engineers spend much time investigating such failures.

Forensic engineering: the investigation of failures. Thomas Telford, London, 2001.

Crocker [1990] makes an analogy between building failure and people. Crocker states: "People deteriorate as part of the natural processes of aging, and so do buildings. People also deteriorate, either temporarily or permanently, as a consequence of some malfunction, and again, so do buildings."

A stable building is in equilibrium such that collapse will not occur. Webster's defines equilibrium as "a balance; a state of even balance; a state in which opposing forces or tendencies neutralize each other." The concept of stability, according to Webster's, is the ability to establish, maintain or regulate the equilibrium of can be described by considering equilibrium of a rigid ball in various positions depicted below:

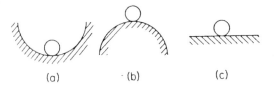

(a)　　　(b)　　　(c)

FIGURE 1

Although the ball is in equilibrium in each of the three positions, if the ball in (a) is displaced slightly from its original position of equilibrium it will return to its original position. Thus it is in a state of stable equilibrium. Yet the ball as shown in (b) displaced and continued to move farther away from its original equilibrium position. This condition is precarious and is called "unstable equilibrium". The ball shown in (c) illustrates yet another type of equilibrium where the ball after being displaced neither returns to its original position nor continues to move farther away instead it remains at the position to which the force has displaced it. This behavior is referred to as neutral equilibrium. This concept is essential in analyzing vertical (gravity) loads in which members might fail due to buckling; however, many structures are often provided with redundant members that prevent collapse. Although there may be intentional redundant elements in a building, there are also unintentional redundant factors such as gypsum wallboard, cladding, interior partitions and other elements that provide support both vertical and horizontal. Where these redundant factors are not present, progressive collapse may ensue due to the collapse of an individual element such as a non-redundant roof framing arrangement where one truss fails, thus overloading the next adjacent truss causing a domino effect literally "un-zipping" the roof. According to McKaig [1962] "Usually buildings fail through men's ignorance, carelessness, or greed." Crocker [1990] listed six causes of failure that include:

1. Natural phenomena such as storms, resulting from damage from floods, exceptionally high winds, lightning, earthquakes.

2. Design errors.

3. Workmanship errors and faulty materials.

4. Procedural errors.

5. Failure to maintain properly.

6. Abuse or misuse of the building.

Blockley [1980] also listed causes of structural failure including

Limit states

Overload:	geophysical. dead, wind, earthquake. etc., manmade. imposed. etc.
Under-strength:	structure, materials, instability
Movement:	foundation settlement. creep, shrinkage. etc.
Deterioration:	cracking, fatigue, corrosion, erosion, etc.

Random hazards

Fire

Floods

Explosions:	accidental, sabotage

Earthquake

Vehicle impact

Human-based errors

Design error:	mistake, misunderstanding of structural behavior
Construction error	mistake, bad practice, poor communications

When the demand on the structure exceeds its capacity collapse will occur. The lack of redundant factors in the structure will determine the extent of collapse that is. if it is partial or complete. For example, when one roof truss fails the load is transferred to the adjacent truss, overloading it and causing it to fail, thus overloading the next adjacent truss. etc. causing a chain reaction or "un-zippering" of the roof structure as presented by Simons [1999] in Figure 2 below.

Although faulty design and faulty construction are sources of problems. The structure can also be overloaded during construction with loads not anticipated by the designer. Modifications of the structure after its initial construction can create conditions conducive to the collapse similar to that described by Simons [1997] who discusses a riding arena building that was originally a battleship dry dock cover that had been relocated to a different site.

FIGURE 2

When re-erected on the new site the columns had been moved outboard approximately 2' from their original location thus introducing large eccentricities into the heels of these long span timber trusses causing large prying forces in the heels of the trusses.

APPLICABILITY OF BUILDING CODES

The 1997 Uniform Code for the Abatement of Dangerous Buildings as presented by The International Conference of Building defines dangerous building.   In Section 302 there are several categories, however in Section 302-8 "Whenever the building or structure or any portion thereof, because of (i) dilapidation, deterioration or decay; (ii) faulty construction; (iii) the removal, movement, or instability of any portion of the ground necessary for the purpose of supporting such building; (iv) the deterioration, decay or inadequacy of its foundation; or (v) any other cause, it is likely to partially or completely collapse." Section 302-11 states: "Whenever the building or structure, exclusive of the foundation, shows 33% or more damage or deterioration of its supporting member or members, or 50% damage or deterioration of its non-supporting members, enclosing or outside walls or coverings." Section 302-14 states: "Whenever any building or structure which whether or not is erected in accordance with all applicable laws, ordinances, has any non-supporting part, member of portion less than 50% or in any supporting part, member of portion less than 66% of the (i) strength; (ii) fire resisting qualities or characteristics; (iii) weather resisting qualities or characteristics required by law in the case of a newly constructed building of like area, height and occupancy in the same location." In essence, these requirements indicate that if the load bearing elements of the building have been reduced 33% by deterioration then the building is deemed to be unsafe and dangerous.

DECAY

According to the American Institute of Timber Construction [1985], decay of wood is caused by low forms of plant life (fungi) that develop and grow from microscopic spores which are present wherever wood is used. The fungi convert wood substance into food. If deprived of any one of the four essentials of life (food, air, moisture, or favorable temperature), decay grown is prevented or stopped and the wood remains sound, retaining its existing strength with no further deterioration. Wood will not be attacked by fungi if it is submerged in water, thereby excluding air, kept continuously below 20% moisture or maintained at temperatures below freezing or above 100° F.   The early stages of decay are often accompanied by discoloration of the wood, which is more evident on freshly exposed surfaces of unseasoned wood than on dry wood. However, many fungi produce early stages of decay that are similar in color to that of normal wood or they give the wood a water-soaked appearance. Further stages of decay are easy to recognize because the wood has undergone definite changes in color and texture. Later stages, all decay fungi seriously reduce the strength of the wood and also its fire resistance. Brown, crumbly, rot, in the dry condition, is sometimes called dry rot, but the term is incorrect because the decay fungi must have some source of moisture for development even though the wood may have subsequently become dry. The American Society of Engineers Design Guide and Commentary on Wood Structures also state that moisture content of 20% or higher the supply of water is adequate for fungi to actively consume either the cellulose or lignin factions of wood. The fungi in infected wood will lie dormant during dry periods and become active when wetting occurs. Thus, previously it took many years to decay the wood to a state where collapse would occur. A 10" wide by 24" deep wood beam in an 80-year old building that is partially collapsed is shown in Figure 3.

FIGURE 3

The effects of decay have been known for ages. According to Sack and Aune [1989] 10th Century Norwegian builders constructed small wood stave churches using vertical timbers imbedded in the ground that have survived for more than 800 years. The researcher stated that "the foundation raft and the sill beams were the most susceptible to decay, but we found only one that was in advanced stages of deterioration." The presence of drainage holes along the base of the sills and the good rock foundation seemed to insure a long, useful life even for this vulnerable component. In the United States, Conwill [1989] discussed covered wooden bridges that hold a place in the idyllic folk image of American history. Although it was theorized that these bridges were covered to reassure skittish horses or to protect the travelers, when builder Timothy Palmer constructed the first known covered bridge in America in Philadelphia in 1805, he gave durability as the reason for covering the bridge. These bridges whose wood trusses were exposed to the weather would last only ten or twelve years, yet Palmer stated that the covering would add greatly to the life; although, he admitted, "there are some that say I argue much against my own interest." Conwill [1989] presented that clear proof of the difference the covering makes is offered by certain bridges which have had some portion poorly maintained while the rest is well maintained.

SOURCES OF WATER INFILTRATION

Sources of water infiltration that will promote decay include sources from the interior and the exterior. The interior sources are caused by condensation of water vapor that is being transmitted from the interior of the building. As the temperature of the air containing a given amount of water vapor falls, the relative humidity rises until the saturation or dew point temperature is reached and some of the vapor is condensed. This condensation may occur on cold, interior wall surfaces and be apparent; however, water vapor will pass through the interior wall surface and since the interior of the wall becomes progressively colder as the outside is approached a temperature may be reached at which the vapor will condense within the wall. The conditions which are favorable for this action are an inside humidity greater than 35% and long continued cold outside temperature. The amount of water which condenses gradually increases and, if the temperature where the water collects is below freezing, ice will form. If insulation is present, it will accumulate water and ice and largely lose its effectiveness. Ice will melt when the outside temperature rises sufficiently.

Condensation may also occur in insulation over the ceiling of a top story, or on the underside of the roof sheathing particularly around the points of protruding roofing nails. The condensation which forms in the walls, ceilings or roof construction may come through the finished wall and ceiling surfaces and cause staining and damage the building in other ways. Condensation distress is sometimes wrongly attributed to leaking walls and roofs. In current wood-framed construction, as currently is provided in accordance with the energy conservation guidelines, the wall may be so tightly sealed with water resistant barriers at the exterior and vapor barriers (retarders) at the interior that evaporation of any moisture that accumulates in the wall occurs very slowly.

According to Carper [1989], the most costly recurring performance problems are those associated with building envelope performance. The sources of water entry from the exterior can include the roof and the exterior cladding, i.e., wood siding, stucco, synthetic stucco, etc. One of the most problematic instances of water infiltration through exterior walls are either through or around penetrations in the walls. Many of the penetrations in buildings are provided to accommodate windows. According to Nicastro and Simons [1997], water infiltration is the source of 50% to 75% of construction litigation claims. Nicastro and Simons state that "water infiltration through exterior walls often appears as window leaks; however the windows do not leak in many cases but more often they create a penetration that interrupts the exterior waterproofing systems.

Hollow window frames can also act as conduits channeling water from remote locations to the window sills or heads. In residential and light commercial construction self-flashing window units are common. These prefabricated windows come complete with a flange around the perimeter for nailing and can be used as a flashing "tie-in" to the weather resistant barrier. Designers often indicate that the windows should be installed according to the manufacturer's instructions, but the manufacturers usually do not provide instructions for interfacing with the weather resistant barrier. The result is an epidemic of improper installations. When these window units are not properly installed and leakage develops around the window, the tendency is to apply copious amounts of sealant over all possible external openings. This application of sealant is unfortunate, because the flanges, flashings and underlayment act as the secondary (last) barrier but they must weep collected water through the wall assembly at some location. Thus inappropriate remedial sealant application can block the discharge of water from the wall system. Although no industry standard has been established for the proper flashing of windows, Section 1402.2 of the Uniform Building Code [1997] requires that "exterior openings exposed to the weather shall be flashed in such a manner as to make them weatherproof". A recommended practice of window flashing is shown in Figure 4.

FIGURE 4

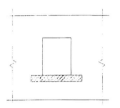

STEP 1. Attach Sill Strips Level with Sill and Extend out beyond Opening.

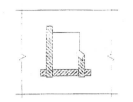

STEP 2. Attach Jamb Strips Even with Opening Sides and Extend above Head and below Sill Strip.

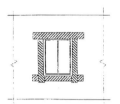

STEP 3. Place Window into Opening with Nailing Flanges over and Sealed to the Sill and Jamb Strips with a Compatible Sealant, then Install Head Strip.

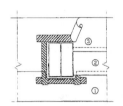

STEP 4. Install Underlayment Beginning at the Bottom with Layer 1, such that It Lies Beneath the Sill Strip and Continuing with the Additional Overlapping Layers 2 and 3, up the Wall, over the Jamb and Head Strips.

When nailing flanges are not present on the windows the weather resistant barrier must be integrated to the opening around the windows to prevent water infiltration. Often times the design of the system calls for the window to be recessed within the wall opening which results in a horizontal surface that cannot be adequately flashed with a weather resistant barrier. At these locations a typical asphalt impregnated felt building paper is insufficient to resist water infiltration and thus the opening must be flashed with either plastic or metal flashings. Essential to the installation is the sealing of the joints at the corners of the flashings. Leakage at the window location cannot only come from the interface between the window and the exterior wall siding; however it can come from the window itself. It is possible for the window to be overwhelmed by wind-driven rain such that the water passes from the exterior to the interior and onto the interior finishes and possibly down into the wall cavity. However, more common leakage from the frame of the window takes place. Aluminum windows that are fastened with screws between the jamb and the sill often leak at this interface. Vinyl windows if properly installed are subjected to leaking from building distortion. Improper hand trans-

portation can result in cracks in the vinyl frame and vinyl windows that have intermediate vertical members (mullions) providing structural strength have been discovered to have been fastened with screws that are not sealed. These screws penetrate the sill track of the window which can allow water to drain into the exterior wall cavity (refer to Figure 5).

FIGURE 5

The vinyl windows which have been assembled from two or more individual units to form a larger "gang" window, the propensity to leak at the vinyl extrusion or meeting rail that is used to connect the adjoining windows. These extrusions can be either vertical or horizontal. Wood windows are not immune to leakage either. Although the frame may be square and well joined when constructed, subsequent expansion and contraction of the wood window from temperature, shrinkage and swelling of the wood window from changes in moisture content will cause separations to form at the mitered joints at the corners of the windows. This can direct water from the window into the exterior wall cavity (refer to Figure 6).

FIGURE 6

Other sources of water infiltration include: diversion of water from the roof behind the wall system that is inadequately protected without diverter (kick-out) flashings (refer to Figure 7).

Buildings that have balconies or wood decks are often subjected to water infiltration through mitered joints at the wood caps above the guard rail walls of the decks or the lack of an adequately counter-flashed joint where the guard rail wall intersects the main field of the exterior wall of the building.

FIGURE 7

DESIGN CONCEPT

Although several technical areas of water infiltration have been discussed, the general change in building design within the last twenty years has resulted in increased demands upon the building for weather resistivity and resistance to decay. These changes include the following:

1. The absence of roof overhangs creating a more monolithic "modern" appearance of the building.

2. The advent of barrier siding systems relying upon the exterior surface of the building to restrict all water entry.

3. Reliance on sealants.

4. The increase in restrictive energy requirements increasing insulation in the walls and restricting air infiltration.

5. The use of new man-made wood cladding materials that have a great propensity to expand (swell) when there is increase in moisture content.

6. The use of various types of windows with no basic standard for flashing.

7. The use of new growth lumber that has a greater propensity to decay than the denser old growth lumber.

DEVELOPMENT CULTURE

Last but not least, is the discussion of the development culture in the construction of multi-family/condominium and apartment type buildings. It has been observed that developers initially hire architects to develop sufficient drawings in which financing can be obtained. Building departments review the drawings on a basis of life safety and typically are not concerned with the weatherproofing details of the building. These details that are necessary to prevent water infiltration are omitted by the architect and developer due to cost. In addition, the architect's services are normally not retained during the construction phase and thus reliance upon weatherproofing of the building is placed upon the shoulders of the individual subcontractors including:

- The framer who erects the wood framing and installs the windows.
- The exterior cladding contractor who installs the weather resistant barrier.
- The roofer.

When the building begins leaking, the typical dispute arises that the framer did not provide the proper flashings around the windows such that the wall cladding contractor (sider) could interface them properly with the weather resistant barrier, and the wall cladding contractor and the roofer point at each other's responsibility as to who was to install the diverter flashings to prevent water from the roof entering behind the exterior wall cladding. As the general contractor blames the subcontractors, the subcontractors in turn blame the general contractor for a lack of coordination. Typically, in the Puget Sound Region, the architect is normally not included in this dispute due to their purposeful omission of the details necessary for a properly weatherproofed building. Thus this responsibility either falls back onto the developer or general contractor.

CONCLUDING REMARKS

Imminent collapse due to hidden decay is a potentially widespread dilemma in many areas of the world where conditions are conducive to decay formation. The advent of newer energy efficient construction that limits the dissipation of moisture from the structure has accelerated the decay process such that deterioration of structural members which previously took decades or even centuries to take place now can occur within a few years. If the current construction and development practices are not altered, it is likely that prescriptive requirements will have to be implemented by governmental bodies in order to protect the safety of the public.

REFERENCES

American Institute of Timber Construction, *Timber Construction Manual, 3rd Edition,* John Wiley & Sons, New York, 1985.

American Society of Civil Engineers, *A Design Guide and Commentary Wood Structures,* ASCE, New York, 1975.

American Society of Civil Engineers, *Evaluation and Upgrading of Wood Structures: Case Studies,* ACSE, New York, 1986.

Blockley, D.I., *The Nature of Structural Design and Safety.* Chichester, West Sussex: Ellis Horwood Limited, 1980.

Carper, Kenneth L. *Forensic Engineering.* Elsevier Science Publishing Co., New York, 1989.

Conwill, Joseph D. *The Covered Wooden Truss Bridge, Classic Wood Structure,* ASCE, New York, 1989.

Crocker, A. *Building Failures Recovering the Cost,* London BSP Professional Books, 1990

Gordon, J. *Structures; Or Why Things Don't Fall Down,* Penguin Books, New York, 1978.

Huntington, Whitney Clark, *Building Construction,* John Wiley and Sons, Inc., New York, 1950.

International Council of Building Officials, *Uniform Code for the Abatement of Dangerous Buildings,* ICBO, Whittier, CA , 1997.

McKaig, Thomas K. *Building Failures: Case Studies in Construction and Design.* New York: McGraw-Hill Book Company, 1962.

Nicastro, David H., Simons, Kenneth B., *Fenestration Frustration: Failure Mechanisms in Business Construction,* ASCE, New York, 1997.

Sack, Ronald L. and Aune, Petter. *The Norwegian Stave Church A Legacy in Wood, Classic Wood Structures,* ASCE, New York, 1989.

Simons, Kenneth B. *Challenges Created by the Hierarchy in Facility Rehabilitation Projects,* Proceedings of the 1999 Structures Congress, Reston, VA, American Society of Civil Engineers, 1999, pp. 885-888.

Simons, Kenneth B. *Should I Notify the Building Department? The Dilemma of the Forensic Engineer,* Proceedings of the First Forensic Congress, Reston, VA, American Society of Civil Engineers, 1997.

Webster's *New Compact Format Dictionary,* Pro Sales, Inc., New York, 1986.

Architects:
What can they learn from Forensic Engineers?

KENNETH L. CARPER, Professor, Reg. Architect, Member ASCE
Editor-in-Chief, ASCE *Journal of Performance of Constructed Facilities*
School of Architecture & Construction Management, College of Engineering & Architecture,
Washington State University, Pullman, WA 99164-2220, USA

Architectural publications tend to focus on success stories and creative new developments in the profession. Articles on failures and performance deficiencies are very rare in these publications. The Technical Council on Forensic Engineering (TCFE) of the American Society of Civil Engineers (ASCE), recognizing the need for interdisciplinary discussion of the causes and costs of failures, attempted to involve the leading architectural organizations in TCFE activities. These efforts by ASCE have met with limited success. Indeed, in some cases, the overtures have been met with outright rejection. This paper describes a new publication and continuing education course, *Why Buildings Fail*, specifically prepared for practicing architects by the National Council of Architectural Registration Boards (NCARB). The project may be the first initiative undertaken by a U.S. architectural organization specifically aimed at incorporating lessons from forensic engineering into architectural practice.

BACKGROUND
The Technical Council on Forensic Engineering (TCFE) was established by the American Society of Civil Engineers (ASCE) in 1982. One early activity of the Council was to undertake an inventory of failure information dissemination activities in the construction industry [Carper 1987]. This study found that, while there were a number of ongoing efforts to discuss forensic engineering topics, most were of limited scope and focused on specialized disciplines. The need was identified for a journal that would provide a forum for interdisciplinary discussion on the causes and costs of failures. Furthermore, it was decided that this journal, if it were to be successful in reducing the severity and frequency of failures, should include discussion of not only the technical causes of failure, but also the human errors that contribute to deficient performance of constructed facilities. Failures are seldom the result of errors made by a single individual. Failures usually involve multiple errors and flawed decision processes, including communication deficiencies and confusion regarding roles and responsibilities of the various parties that make up the project delivery team.

An international editorial board comprised of professionals representing a wide variety of design and construction disciplines was formed to create the new ASCE *Journal of Performance of Constructed Facilities*. The first issue of the journal appeared in 1987; this year will be the 15th successful year of publication.

Forensic engineering: the investigation of failures. Thomas Telford, London, 2001.

From the beginning, an intense effort was undertaken to involve the American Institute of Architects (AIA) as a cosponsoring organization for the journal. No financial support was requested from the AIA. It was simply ASCE's intention to designate the AIA as a cosponsoring organization, along with several other professional societies that are listed at the front of the journal, including the National Society of Professional Engineers (NSPE), Professional Engineers in Private Practice (PEPP) and the Architecture and Engineering Performance Information Center (AEPIC). Encouraging discussions with leading architects and meetings at the AIA corporate office suggested, right up until the time of publication, that AIA would formally participate as a cosponsor. However, just before publication of the first issue, the legal counsel at the corporate AIA office rejected the invitation.

Her reason was simply stated: "Architects don't read."

Furthermore, she opined that architects must be protected from information that could be damaging to their position in litigation. Her concern was that an architect might be confronted in the courtroom after using a detail that had led to repetitive problems in the past. Should such information have been presented in the journal, the architect might be queried as to why the architect had not "read his (or her) own journals." Thus, the decision was made that the good name of the American Institute of Architects should not be connected with any activity focused on learning from failures. Despite an open invitation and the continued support of many of its individual members, the AIA organization has expressed no further official interest in the journal since this decision was made.

The typical architectural journal is a beautiful publication. Glossy photographs and inspiring prose glorify the latest achievements in the field. Seldom are projects revisited for information on their actual performance, and almost never are performance deficiencies and failures discussed in these publications. Only a handful of articles on a few of the most dramatic structural collapse events have appeared in the leading architectural journals. The typical articles may encourage creativity and inform practitioners of new materials and methods. However, many educational opportunities have been lost as a result of the conscious decision to avoid open discussion of failure-related information in the pages of these influential journals.

One national architectural organization, however, has taken a positive initiative in the direction of learning from forensic engineering case studies. The National Council of Architectural Registration Boards (NCARB) has recently begun to address the need for continuing professional education for architects. NCARB's Professional Development Program is providing resources to enhance the technical competence of practicing architects. In particular relevance to the subject of this paper, NCARB has made a valuable contribution by establishing a new course on building failures, specifically prepared for the architectural profession.

NCARB PROFESSIONAL DEVELOPMENT PROGRAM

The National Council of Architectural Registration Boards is a non-profit federation of 55 state and territorial architectural registration boards in the United States. The mission of NCARB is to develop professional standards for architects. Services provided by NCARB support a wide range of important activities including: intern development programs; architectural registration examinations; reciprocity arrangements for registration across state borders; national certification; and continuing professional education.

Established in 1993, NCARB's Professional Development Program and its monograph series were created to address the need for continuing education and professional development verification for architects. An increasing number of states now mandate continuing education for periodic renewal of architectural registration. All U.S. jurisdictions accept NCARB Professional Development Program monographs for compliance with state laws. NCARB is also an approved provider in the Continuing Education System of the American Institute of Architects (AIA).

Each monograph published by NCARB includes a quiz on the subject. After self-study of the monograph, participants complete the quiz and return it to NCARB for scoring and continuing education verification. Participants retain the monograph for future reference in practice. To date, NCARB has published ten monographs, including such titles as *Seismic Mitigation*, *Low-Slope Roofing*, *Wind Forces*, *Sustainable Design*, *Energy-Conscious Architecture*, and *Fire Safety in Buildings*. These topics represent NCARB's commitment to provide educational resources that will enhance the technical competence of the architectural practitioner. This paper reviews the latest NCARB monograph, *Why Buildings Fail* [NCARB 2001].

Further information regarding the National Council of Architectural Registration Boards and its Professional Development Program can be found at the NCARB web site: http://www.ncarb.org.

WHY BUILDINGS FAIL: NCARB CONTINUING EDUCATION COURSE
The most recent Professional Development course available from NCARB is based on the monograph *Why Buildings Fail* [NCARB 2001].

Numerous case studies are presented in the monograph to illustrate the sources of building failures. Throughout the volume, "failure" is defined as "performance that does not meet expectations." Thus, in addition to examples of structural collapse, there are also case studies of serviceability problems, cladding failures, premature deterioration, mechanical/electrical system failures and other functional deficiencies. Examples are drawn from the ASCE *Journal of Performance of Constructed Facilities* and a number of forensic engineering references [Kaminetzky 1991, Shuirman and Slosson 1992, Feld and Carper 1997, Carper 2001, and many others]. The 119-page monograph is illustrated with line drawings and photographs. The writing style is specifically addressed to practicing architects.

Case studies are presented with the intent of clarifying the underlying causes of failures and performance deficiencies. The goal is to reduce the repetition of failures through studying their sources, not to criticize the unfortunate individuals involved with the specific cases.

The monograph is organized in such a way as to illustrate that errors can be introduced at any phase of a project, from initial site selection and programming, through design development, construction, and operation. While the architect's responsibilities and authorities are limited, it is nonetheless important for the architect to have an understanding of the many sources of error that can eventually contribute to a project's success or failure. The volume includes discussion of failure-avoidance strategies, failure litigation and alternative dispute resolution techniques. An extensive reference list is provided for further study.

CONTENTS OF THE MONOGRAPH

Case study examples are presented and discussed in the following order:

- **Forces and destructive agents: natural and human-caused hazards**
  Buildings: static objects in a dynamic world; Gravity; Wind; Earthquake; Flood; Fire; Unusual loads: blast, vibration and collision; Time-related deterioration; Other factors.

- **Fundamental errors at the outset of a project**
  Flaws in the basic concept; Site selection and site development errors; Programming and project definition deficiencies; Pathology and teratology.

- **Errors during the design phase**
  Design challenge: anticipation of failure; Failure to meet applicable codes and standards; Engineering design errors: structural, mechanical/electrical, other technical disciplines; Material selection and detailing errors; Inadequate concern for durability and maintainability; Incomplete or inconsistent project documents.

- **Errors during the construction phase**
  Communication, coordination and inspection; Construction site safety; Excavation accidents; Improper sequencing of construction; Inadequate temporary support for unstable assemblies; Excessive construction loading; Temporary weaknesses: material and connection vulnerabilities.

- **Operational errors**
  Anticipating post-occupancy problems; Misuse and abuse; Operational problems with sophisticated equipment; Alterations, renovations and change in use; Deficient maintenance.

- **Conclusion**
  Failure-avoidance strategies; Failure litigation and dispute resolution.

- **Appendix: lessons from failures**
- **Bibliography**
- **Quiz**

SUMMARY: DO ARCHITECTS READ?

Practicing architects, like practicing professional engineers, constantly face time pressures in their work. A British architect sums up the challenges of architectural practice:

> *I am an architect and like all architects I know nothing about anything, but I have to make decisions about everything. That is a terrible burden. Architects are often condemned for the foolish things they do and they do many of them. But if you look at it from the architect's point of view, he starts with an empty site, with a blank piece of paper, an equation with thousands of unknowns and relatively few knowns. He has to solve the equation by Friday and make it beautiful too. With a problem like that, designers have to guess the unknown on the basis of the best information available.* [Hutton 1996]

Professional engineering societies are challenged to develop efficient communication techniques for the delivery of relevant information to practicing engineers. Practicing architects have similar needs.

Architectural societies in the United States are responding to the continuing education needs of their members. The success of NCARB's Professional Development Program shows that architects, like engineers, will read—but only if the information is presented in a way that is relevant to practice.

The Technical Council on Forensic Engineering (TCFE) of the American Society of Civil Engineers continues to produce resources that are of interest to architects. Several architects are contributors to these activities. There are architects on the editorial board for the ASCE *Journal of Performance of Constructed Facilities*. A number of architects subscribe to the journal, and architects have contributed articles on some very interesting case studies. Indeed, the "Outstanding Paper Award" was presented to an architect last year for an article on a brick cladding failure [Anderson 1999].

Contrary to the opinion expressed earlier by the legal counsel for the American Institute of Architects, many architects not only read but are also profoundly receptive to information resulting from forensic investigations. Members of TCFE have regularly been welcomed for presentations to local chapter meetings of the American Institute of Architects. For a number of years, TCFE's Committee on Dissemination of Failure Information contributed a regular column in *Specifier*, the journal of the Construction Specifications Institute (CSI). The column, titled "Critical Details," focused attention each month on a specific architectural detail that had caused problems. Architects have indicated that the "Critical Details" column was the first item they looked for in each new issue of *Specifier*. These articles have been collected by ASCE and published in a single volume [Nicastro 1997].

Another ASCE publication that should be useful to architects is a product of TCFE's Committee on Education [Shepherd and Frost 1995]. This monograph is a brief summary of approximately 50 case studies involving structural, foundation and geoenvironmental failures. Ongoing TCFE activities include the development of an easily accessed case study web site [Rens, Clark and Knott 2000] and preparation of a series of failure vignettes on CD-ROM [Zickel 2000].

While the leading architectural publications may continue to avoid discussion of failures, it is gratifying to see an increasing interest level among individual practitioners. And it is especially encouraging to observe the initiative undertaken by the National Council of Architectural Registration Boards in introducing these topics to their membership by way of an enlightened Professional Development Program.

REFERENCES

Anderson, Lindsay M., "Spalling Brick: Material, Design, or Construction Problem?" *Journal of Performance of Constructed Facilities*, Vol. 13, No. 4, American Society of Civil Engineers, Reston, VA, November, 1999.

Carper, Kenneth L., "Failure Information: Dissemination Strategies," *Journal of Performance of Constructed Facilities*, Vol. 1, No. 1, American Society of Civil Engineers, Reston, VA, February, 1987.

Carper, Kenneth L., editor, *Forensic Engineering (2nd edition)*, CRC Press, Boca Raton, FL, 2001.

Feld, Jacob and Carper, Kenneth L., *Construction Failure (Second edition)*, John Wiley & Sons, Inc., New York, NY, 1997.

Hutton, Geoffrey, "Introduction," in *Allergy Problems in Buildings*, Singh, Jagjit and Walker, Bryan, editors, Quay Books, Mark Allen Publishing, Ltd., Wiltshire, UK, 1996.

Kaminetzky, Dov, *Design and Construction Failures: Lessons from Forensic Investigations*, McGraw-Hill, Inc., New York, NY, 1991.

NCARB, *Why Buildings Fail*, by Kenneth L. Carper, National Council of Architectural Registration Boards, Washington, DC, 2001 (ISBN 0-941575-36-5).

Nicastro, David N., *Failure Mechanisms in Building Construction*, American Society of Civil Engineers, Reston, VA, 1997.

Rens, Kevin L., Clark, Melissa, and Knott, Albert W., "A Failure Analysis Case Study Information Disseminator," *Journal of Performance of Constructed Facilities*, Vol. 14, No. 3, American Society of Civil Engineers, Reston, VA, August, 2000.

Shepherd, Robin and Frost, J. David, editors, *Failures in Civil Engineering: Structural, Foundation and Geoenvironmental Case Studies*, Committee on Education, Technical Council on Forensic Engineering, American Society of Civil Engineers, Reston, VA, 1995.

Shuirman, Gerard and Slosson, James E., *Forensic Engineering: Case Histories for Civil Engineers and Geologists*, Academic Press, Inc., San Diego, CA, 1992.

Zickel, Lewis L., "Failure Vignettes for Teachers," *Proceedings of the Second Congress: Forensic Engineering*, Rens, Kevin L., Rendon-Herrero, Oswald, and Bosela, Paul A., editors, San Juan, Puerto Rico, American Society of Civil Engineers, Reston, VA, 2000.

Learning from failures – experience, achievements and prospects

M. F. DRDÁCKÝ
Director, Institute of Theoretical and Applied Mechanics of the Academy of Sciences of the Czech Republic, Prague, Czech Republic

BUILDING PERFORMANCE DATA – THEIR COLLECTION, EVALUATION AND UTILIZATION
Background in the Czech Republic
Every year hundreds of events take place all over the world resulting in qualitative and quantitative objective data on the building performance. The data is primarily the result of experiments with large scale models, of field investigations of real buildings or structures and of building failures and collapses.

For its importance, the problem became part of the former Czechoslovak plan of scientific research. Primarily concerning the operational performance of building structures with particular reference to design of an information system using data on structural failures and collapses considered as non-planned experiments. The research programme was controlled by the Institute of Theoretical and Applied Mechanics (ITAM) of the former Czechoslovak Academy of Sciences under the leadership of the author. Participants included the Czech University of Technology in Prague, the University of Technology in Brno, the State Research Institute of Materials Protection in Prague and the Building Technical and Testing Institute in Prague.

The Czech National Information System on Building Failures
In 1980 the ITAM was asked by the Architecture and Engineering Performance Information Centre (AEPIC) of the University of Maryland to cooperate in the collection, evaluation and utilization of data on building failures. For this reason, the basic principles of organization and function of the AEPIC information centre [1] have become the model of the developed Czechoslovak system. Independently, the same year the catalogue of data from Czechoslovak building failures was issued, based mostly on expert archival information from the Building Technical and Testing Institute who have been monitoring defects and failures in civil engineering constructions since 1975. In the field of corrosion, research of corrosive damage to weathering steel in buildings has been organized since 1977 on international basis in cooperation with the State Research Institute of Materials Protection and the Corrosion Institute in Stockholm. From background on these projects the Czechoslovak approach to the utilization of data on structural failures has been formed.

Collection of Relevant Data
In the general variant of the system data cover all aspects of problems arising in civil engineering, concerning building structures, the mechanical and electrical engineering equipment of buildings, building physics, economy, environmental comfort, thermal and

Forensic engineering: the investigation of failures. Thomas Telford, London, 2001.

acoustic characteristics of buildings as well as appearance, aesthetics and functional performance of buildings and structures.

In a limited version, attention is devoted to one aspect of the system which covers only the problems of mechanical performance of structures, including their relationship to the more general behaviour of the before mentioned fields. However, the data are not limited to that obtained from structural failures only. All collected data refer to materials, elements, systems, technological processes and construction management without including any personal data or other facts, which might impair the reputation of firms or personnel.

The monitored data are supplied by three groups of information sources:
1. experimental investigations of full scale structures or their large scale models,
2. loading tests of structures,
3. structural failures.

Also a possible utilization of information obtained by other technical surveys, periodic inspections, etc. is envisaged. For example, an increased attention has been paid to behaviour of historic materials and structures and the data have been collected in a special database.

The data are obtained first from research and testing institutes, technical universities and technical medium-grade schools, further from industrial enterprises, design organizations, insurance firms as well as from individual experts and forensic engineering.

Data Processing
The data base uses data described in literature, published and unpublished reports, expert's accounts, etc., also data contained in special reports prepared in connection with investigations of individual cases of failure or other manifestations of structural behaviour.

The data sheet is utilized for data collection from all three before mentioned information sources. It contains three data sets:
I.   Introductory citation part
II. Basic technical data on the event or set of events concerned
III. Supplementary data on the author of the data sheet, on the filling of the source report, notes, etc.

The data sheet is compatible with the record in the Citation File of the database of the AEPIC International Centre particularly in the introductory part. The basic data set is more detailed and, moreover, classified according to information source group. An example of the structure of this set, comprising of basic technical data for the description of structural failures. This is shown in Table 1, where the filled-up form is presented [2].

Only the data from the data sheet are computerized. In the data sheet both verbal expressions and numerical codes can be used, particularly for the description of such general characteristics, as building type, type of structure, material, etc. The thesaurus of the database is developing continuously in the course of data storage with the maximum use of coded terminology. The methodology of the structure of numerical codes is based on standard codes used in the standard classifications of buildings, construction operations, etc.

Analysis of Structural Failures and Collapses - a Czech Experience
The generalization of results concerning the behaviour of building structures and their failures can only be made on the basis of a sufficiently representative random sampling. So far the most representative partial analysis of the causes of failures prepared in the former Czechoslovakia is based on the data of expert's reports from the Building Technical and Testing Institute [3]. It summarizes the investigations of 570 buildings and structures carried out in 1986 / 1988, 62 % of which presented some failure. This set does not include monothematic investigations of specific, and for that period typical, problems covering hundreds of buildings and structures, such as the investigation of defects and failures of flat roofs, the research of R. C. tramway track panels failures etc., as well as investigations into failures of historic buildings and structures. If used, they are considered as a single case of failure only, not to interfere with the random selection. Another study [4] is based on one hundred cases of steel structure failures recorded by the Czechoslovak State Insurance Company.

Table 1. Example of a filled-out data sheet

TITLE	Small and medium size flagpoles in Prague – Strahov
AUTHORS	Lapka J., Knotková D., Jansenová I.
SOURCE	Expert's opinions from years 1987-88, Report No.40/88
VOLUME	
NUMBER	
PAGES	
PUBLISHER	State Research Institute of Material Protection
PLACE	Prague
DATE	1988
TYPE OF OBJECT	Tube flagpoles fixed in soil, in asphalt pavement of platforms (small size) and in concrete footings (medium size poles).
LOCATION	Prague – Strahov stadium
CONSTRUCTION	
Beginning	1955
Termination	1955-58
DATUM OF EVENT	1987, 1988
AGE OF OBJECT	30 years
TYPE OF STRUCTURE	Hollow steel columns with uniform and step-wise variable cross-section (steel grade 11).
CONDITIONS	Town atmosphere, exposed zones in the vicinity of anchoring into soil or footings with humidity retaining and waste cumulation.
DATA OF REPORTED	In areas of about 10 cm below and above surface the
EVENTS	corrosion damages are considerably higher than usual for the town atmosphere, (from 4 to 6 times higher corrosion decrease). This decrease reaches 300 micro-meters on external pole surfaces, 800 micrometers on the contact with concrete footings and from 1000 to 2000 micrometers on the contact with soil.
MAJOR CAUSES	Local corrosion on boundary of different phases, water and waste cumulation.
POSSIBILITIES OF	The small flagpole should be replaced, the medium ones
REPARATION	can be conserved by paintings after careful removal old paints and corrosion products.
REMARKS	Documented by photographs.
REPORTER	Knotková
SOURCE LIBRARY	State Research Institute of Material Protection Prague
DATE OF FILLING	December 1989
DATE OF INSERTING	12/1989
RECORD NUMBER	11

The most interesting conclusion is the result of the analysis of what causes individual failures. It has confirmed the well known fact that every failure is usually the consequence of several causes. From the quoted hundred cases [4] only 35 % failures were due to a single cause. The most frequent single causes of failure are faults of erection (14 %), faults in operation of the completed building (10 %) and faults of loading suppositions (7 %). Faulty erection occurs most frequently in cases of failures due to several causes. In these cases the average number of causes was 2.5, the maximum number being 7.

The causes of failures can be divided into four groups of common characteristics:
1. Ignorance and lack of knowledge
2. Errors and negligence
3. Exceptional situations
4. Operational causes.

Ignorance and Lack of Knowledge
Ignorance is a frequent cause of structural failures. This group includes the shortcomings of standards, codes, regulations, etc. Shortcomings in scientific and engineering knowledge as well as shortcomings in availability of information required for the design of structures. From the case available some 6-8 % can be attributed to this group.

Apart from the cases of objective ignorance there is also a number of subjective ignorance cases. They include e. g. a poor competency of persons entrusted with the design, production, erection and inspection of the structures. Poor qualification and insufficient intellectual standard of people carrying out maintenance or supervising operation. Shortage of qualified, independent control of the quality of design documents, shop drawings and quality of finished structures. Although it is difficult to ascertain accurately the number of causes of this type, they occur approximately in 52 % of all cases.

Errors and Negligence
These causes are frequent in investigated cases. They occur and, often multiply, in the majority of investigated cases. It is difficult to distinguish them from the cases of subjective ignorance. Together with the group latter they account for more than 90 % of all found causes. Moreover, gross errors and gross negligence are the main causes of failures due to a single cause.

Exceptional Situations
This number includes particularly natural disasters (earthquakes, windstorms, etc.), fire explosions, impact of vehicles, some technological and technical factors, etc. In the assessed sample [6] they occurred in 1.5 % of cases.

Operational Causes
This group includes particularly failures due to the termination of the structures life, change of the structure performance conditions (load, environment, etc.) and other shortcomings, often having the character of subjective ignorance and errors and negligence groups. They manifest themselves by insufficient inspection and maintenance, non-compliance with operating conditions, and can result e. g. in excessive corrosion which represents one of the serious causes of collapses of steel structures (about 8 %). It should be noted that the shortcomings of maintenance and inspection are often caused by the attempt to economize. This group may include various "improvement" suggestions, economy of materials, acceleration of the construction processes, etc. once again nearing the group of ignorance.

Utilization of Data on Real Behaviour of Structures

Although shortcomings of scientific knowledge represent only about 1 % of failure causes, the collapses of structures resulted in the development of a number of scientific disciplines and still represent a stimulus influencing research directions in the field of the theory of structures. For instance, the fracture mechanics, aeroelasticity, stability of soil structures, methods of experimental analysis of structures and particularly theory of the stability of slender structures, with examples of historic truss bridge accidents as well as more recent collapses of major steel box girder bridges in the phase of erection, which directed research attention to the shear lag problem in wide stiffened flanges and the stability of thin webs subjected to partial-edge compressive loading in seventies of the last century.

In the structural research, besides the above mentioned *research orientation*, the data on building failures might be utilized for description and estimates of *tendencies of random structural characteristics* as well as for *improvements of mathematical models of structural behaviour*.

The data on building failures have also a very practical importance for improvements in building design, construction, service and maintenance practice when available. Therefore, they are a substantial basis for data banks on structural failures and/or for expert systems on structural performance.

DATA BANKS ON BUILDING FAILURES - THEIR PROBLEMS AND PROSPECTS
Costs of Failures or Why Data Banks on Failures

Why is it generally accepted that there exists a need for gathering, evaluating and disseminating data from building failures? Let us illustrate the answer by data from nineties of the last century. According to the qualified estimates costs of building failures reach about 16 billions of Kc (Czech crowns) a year (1992) with many consequences in life, material and environmental loss [5]. In other European countries building damages generate similar figures, e.g. 14 billions of DEM in Germany (1988), 300 millions of US$ in Finland (1990), 3 billions of FF in France (1992), here only   insurance cases. The actual costs are very likely even higher and this information is quite hard to get, especially when we shall consider the below mentioned approach to the definition of a failure.

L. Faria [6] presented results of international investigations into the "economy" of failures of industrial products and concluded, that costs of failures due to fracture, corrosion, erosion and wear reach about 10 % of the GNP in each country. Further, the failures influence the total cost of products by about 2-25 % (for heavy equipments).

How such costs can be reduced? The failures generally, and the building failures especially, are mostly caused by human errors yielding from insufficient knowledge and bad information of people involved and the consequent low level of management and control in the building industry. Here the lack of knowledge means rather education deficiencies than gaps in general knowledge of physical, natural, etc. processes, taking into account the above results of our Czech experience from failures. Therefore, the better education and information systems are considered to be possible tools for reduction of those costs for long, e.g. [1], [6]. This is valid also for prevention of failures of historic buildings and life prolongation of other object of our built environment. Of course, further research in this field is still necessary for improvement of scientific and professional background.

Data Banks on Building Defects and Failures

General or Specialized Data Banks

It is useful to build the data bases (DB) in two basic categories – general DB on building performance or specialized DB supporting mostly expert systems or collecting data for some other narrow purpose. Each type has advantages and disadvantages.

The general DB include a broad scope of cases, they are typically widely accessible to users and mostly useful for first warning, references etc. Examples include the French *Qualité-Construction* with about more than one hundred thousands of records, the American Failure Analysis Assoc. DB with several millions of records, the German *Schadis* or the Czech *POST* DB, each with several thousands of records.

The specialized DB are frequently connected with environmental problems (Polish DB of environmental defects in building, German *Monufakt* on building damages from pollution), or expert systems. They provide more detailed contents and outputs (full text records, graphic documentation), and their filling by data is usually better organized and easier.

Financing Data Banks and Information Systems

In Europe, there exist several modes of financing DB on building failures. In most developed countries, the **state** feels its duty to support this activity, at least partially, as e.g. in Germany by about 65 %. The decision in German speaking countries was drawn from results of a questionnaire investigations which should answer two questions. *Are there conditions supporting market of building information **without** a state support? Is it desirable to have a fully private building failure information market?* The answers to the both questions were NO and due to several reasons. Some of them state that i) the subjects operating in building industry are convinced that they have enough information, ii) the scope is too broad and must satisfy many subjects of very different interests, iii) the effect of better knowledge is often materialized in other fields, i.e. the investor has benefits from the cheaper and quality buildings, but the costs are covered by designers, iv) the cost of collecting, evaluating and disseminating data are so high, that the resulting costs of information services are not competitive, v) costs of creating a DB are returned very slowly, vi) there still exist effective information technologies with a good market, vii) there seems useful to control, to some extent, the building information by state. Similar approach, i.e. **the state support** is adopted in Norway and Spain.

A more effective financing means occur and there are **special programs of Insurance Companies**. This system is partially utilized in Denmark, USA, (Netherlands), and especially France, where it proves to be very reliable and successful. Funds are secured here by some percentage from an obligatory guarantee insurance, which is destined for investigations and documentation of possible defects or failures. The only disadvantage is that the data are collected as coded and therefore of limited contents.

However, there is one good example of creating a data base on failures by **a private company** and it is the above mentioned DB of FaAA (USA) but as to the author's knowledge, the data are not generally available for open access and they serve to the DB owner for their expert opinions, consulting, research and design activities.

Data acquisition represents another difficult and hardly solvable task at creating any DB on building failures. There exists no apparent good will of specialists and construction companies to cooperate and moreover there are heavy legislative obstacles, (silence of experts and witnesses). Further, some data are not sufficiently reliable and this fact need not be apparent.

Publications, Workshops and Conferences

According to the Czech experience the most successful handling with data from failures is to issue specialized books, series, journals and design sheet, as well as to organize special workshops and conferences. Examples include ASCE *Journal of Performance of Constructed Facilities, Journal of Technology, Law and Insurance*, Series *Building Without Damages* in Germany, annual international conferences as e.g. *Building Pathology* (UK), *Lessons from Structural Failures* (CR) and some other less frequent conferences.

METHODOLOGICAL COMMENTS
On Classification of Failures

According to data on failures the author suggested to study three categories of failures in the building industry fields. The first group is characterized by failures in the interrelationship of building structures and materials with their environments, occupants and contents and it is a subject of **Building Pathology**. In other words it is the study of building and structure diseases or functional and structural changes similarly to the effects well known from the field of living organisms. This approach cover defects or failures induced during the building or structure life by environmental or force actions (like fatigue, damage cumulation, corrosion), material degradation (aging, chemical changes), service conditions (misuse, overloading, ignorance), maintenance errors or negligence, etc. The second group of inborn or innate defects including malformations and monstrosities can be called **Building "Teratology"**, conserving the similarity between structural and human behaviour. This approach involves all defects or failures generated during pre-design, design and construction stages of buildings and structures. Here also architectural conception inadequacy often represents the primary cause of serious difficulties and frequently leads to the design of uneconomic systems. The result need not be the physical failure of the structure. However, in almost all cases a moral or aesthetic damage occurs. These defects are difficult to be defined and at structures sometimes even difficult to be ascertained, the structure not being always visible. It is important that the family of these defects initiates more than 80 % of building or structural failures. The last group of failures can be summarized as **Unexpected Events** and it includes particularly natural disasters, fire, explosions, impact of vehicles and some other undesired results of human activity, (industrial failures). At assessment of damages and residual life of structures subjected to accidents the methods of building pathology can be applied. On the other hand, risk minimization and prevention of failures due to unexpected events call for measures substantially different from those of building pathology problems. The unexpected events represent only about 1.5 % of all cases. Nevertheless, they can bring about immense material damages and heavy injuries or loss of life. [7]

On Definitions, Required Patterns, Expected and Optimum Performance

There are several approaches to failure definitions. In [7] the author suggested one which enlarges usual approaches based on comparison of expectations and reality, as it is provided e.g. by the Czech standard terminology [8], where a failure is defined as critical defect and the defect itself is described as "a deviation of an artefact from a required pattern", or by a more general definition used by Technical Council on Forensic Engineering of the American Society of Civil Engineers [9] in the form: "Failure is an unacceptable difference between expected and observed performance". The suggested definition involves also possible failures in setting up the initial concepts and has a form: "Failure is an unacceptable difference between optimum and observed (or achieved) performance".

In any case, we must be able to measure the "acceptable difference" or critical defects as well as to define expected performance or required patterns. Building defects are defined in a relation to some ideal models or patterns and the models themselves sometimes represent a very roughly idealized reality. The required patterns are mostly pre-scribed through geometrical characteristics, their acceptable deviations, the material characteristics and the structural performance in the course of loading of buildings or structures. Such an approach facilitates the theoretical assessments of structural behaviour and safety. But many structures, especially in nature, which do not comply with any Standard or Code Specifications are sufficiently safe and capable to sustain the loads. Are they defective or even failures? In the civil engineering there we must deal with similar problems when assessing damaged structures.

The suggested definition includes focus on a quality of performance. In the authors' opinion, engineering products – or better artificial products and activities – should exhibit *Optimum Performance* (or *Optimum Quality*; furthermore the first term has been adopted for its broader common meaning). All engineers should be educated and convinced to aim at this philosophy. Little attention to the quality of performance is responsible for environmental pollution and energy crises. The term "optimum performance" has been explained in detail in [7], let us mention here only that optimum performance should incorporate technical as well as non-technical aspects of artefact and the optimisation is understood as the achievement of the maximum degree of utility, which at same time fulfils the condition of a minimum energy input necessary for the realization of the objective, and also preserves the environmental and ecological balance.

On Performance Assessments
When assessing artefact performance several criteria accompanied by suitable evaluation tools are utilized to discover defects of which many result in or are declared as failures.

Structural Response to Loads is the most common and widely used criterion describing civil engineering and architecture works. It is logical, that for structural response the *Safety* is a natural measure of defects. David Blockley [10] defines safety as the "correspondence between a required state of the world and the actual state of the world". This definition is sufficiently general to include also non-technical aspects of structures or buildings. It might replace the above definitions of failure on conditions that the words "correspondence" and "required" are suitably explained in detail, like similar general words in the previous definitions. Nevertheless, in the authors' opinion, the term "safety" is too narrow for the general assessment of buildings or structural performance acceptability. For structural response to loadings, as well as functional response of engineering systems, the safety represents the most valuable evaluation tool. The basic theory is very well established and modern approaches are being developed thoroughly [10]. However, also the appropriate safety should be measured and established to be optimum. The main advantage of safety concept is its possibility to be supported by a well elaborated system of codes, apart from the fact that philosophy bases for some of them should be improved, changed or even developed.

Material Consumption represents a problem very firmly connected to the previous one. It can hardly be accepted as "sound" a structure which is sufficiently safe but made of over-dimensioned elements. Various *Shape Optimisation Methods* can be taken advantage of to solve this question. Of course, for many engineering problems other objective functions than mass minimization are more important. This item is mentioned here because of an immense amount of theoretical works devoted to it and still very rarely utilized in engineering practice.

Energy Consumption approach provides a more profound method of structural optimisation in civil engineering. Let me mention an example from [11], which illustrates the problem. There is shown that a local optimum of a roof truss, bringing about a slight increase of weight by 1.2%, results in a structure with substantially lower height than the global optimum truss. The height reduction represents about 1/3 of the global optimum solution. From this result it follows that the structure which is safe, manufactured with minimum consumed material can bring about an increase of energy consumption during its operation if the inner room is heated, in vain increased stresses in supporting structures due to wind action, etc. A sub-optimum solution, from the weight point of view, may give better results from energy consumption point of view. The same is valid for labour consumption during production or erection of a structure. Solution can be achieved here using ***Cost Optimisation Methods***. Nevertheless, such an approach needs appropriate theoretical background and reliable practical data. Because of its basic influence on structural concepts it calls also for new "codes of practice" as well as for changes in the present system of design codes.

Functional Response is hardly to be evaluated if it is concerned with non-technical aspects, i.e. mainly in architectural performance as e.g. layout scheme and its functional linkage. Here probably the ***Expert Analysis*** and/or ***Users Criticism*** are the tools which can be applied at the performance evaluation. Some problems can be transformed on an energy or cost base, as e.g. insolation or shading defects, which are frequently subjects of litigations. To this group we should include also economic characteristics.

Environmental Response has been attracting more and more attention of the public for the last several years because failures in this field represent a threat for Nature, the mankind included. Again, the technical parameters of environmental influences may be evaluated quite sufficiently. On the other hand, the results of long term expositions of living entities to environmental changes and harms are hardly to be discovered and the relevant detail criteria or objective assessment tools are missing. Moreover, into this family of questions not only the aspects of natural sciences but also of health, social and psychical quality of surrounding environment should be included. The necessary tools for quality assessment have not been fully developed.

Aesthetics, Philosophy and Similar Responses may also indicate substantial defects in architecture and civil engineering works. Their adequate evaluation might be based on ***Expert Criticism*** but any conclusions drawn from individual opinions without the possibility of being supported by "computed" or "measured" evidences are very vulnerable. The theoretical analyses, qualified critical reviews taking into account broad public opinion investigations might be helpful. The damages of copying bad examples of "official" architecture will not probably be a subject to any litigation, but their consequences could be massive and terrible.

CONCLUSIONS

It is obvious that the collection of data obtained by experimental research of structures and buildings, loading tests, structural failures and collapses during their life time, in data bases is of enormous significance. By means of these data it is possible to prevent the initiation of failures, to influence the directions of further research, to improve the knowledge of random characteristics of structures and to solidify the general mathematical models of structural behaviour.

Therefore, it was agreed to build a national information system on building performance in the Czech Republic. Due to economic reasons and difficulties with gathering the relevant data it was decided to continue filling the database of the developed system which is operated in the Institute for Building Information in Prague. Nevertheless, the coordination of state activities as well as private activities in introduction of differential Information systems (NIS, GIS, etc.) with gathering data on structural failures was recommended to be followed and developed on international level.

State authorities should be encouraged to support creation of joint projects with other European countries in the discussed activity. International exchange in this field is vital for successful and valuable data supply. In the Author's opinion, it is necessary in structural behaviour to distinguish between data from failures, loading tests etc. and data from experimental research. In the first case a network based on information from national information centres could exchange data on failures and similar cases, provide case studies and also organize national and international conferences. As far as data on experimental investigations is concerned an international network of cooperating institutions, specialized working groups, or laboratories which are engaged in research of special problems or experimental research programmes could be created. It will be useful to coordinate this activity with the help of international organizations or professional associations. **An European harmonization of basic approaches and methodology is vital for the building of any national information system.**

It is necessary to evaluate best practice in funding information systems of structural failures and establish a joint budgetary scheme for support of the operation and further development of databases on building failures by combination of state and insurance or private funds.

It is important to develop editions of special series of publications and organizing of specialized conferences and workshops.

ACKNOWLEDGEMENT

This work was partially supported by the Ministry of Culture Grant No. PK99P04OPP006 and the EC Project ICA1-CT-2000-70013.

REFERENCES

[1] FitzSimons, N.: Making failures pay, Engineering Performance Information Center Project, Kensington (1980)
[2] Knotková, D.: Defects of steel structures caused by atmospheric corrosion, (unpublished)
[3] Schreiber, V.: Analysis of operational behaviour of selected structural types, (in Czech), unpublished Research Report, Prague, (1988)
[4] Marek, P.: Analysis of damages and failures of steel structures, (in Czech), Czechoslovak State Insurance Company Report, Series "Zabraňujeme škodám", Vol. 23, (1988)
[5] Drdácký, M.: Development conditions for the national information system on building failures, (in Czech), Report for the Institute of Building Information, 15 p., November (1992)
[6] Faria, L.: Training Engineers and Technicians to Fight Against Failures, Proc. of the 4th Int. Conf. "Lessons from Structural Failures" (M. Drdácký - editor), pp. 37-42, Prague, (1994)
[7] Drdácký, M. F., Kratěna, J.: Forensic Practice in the Czech Republic, Proc. of the Second Congress „Forensic Engineering" (K. L. Rens, O. Rendon-Herrero, P. A. Bosela – editors), ISBN 0-7844-0482-8, ASCE, Reston, Virginia, pp. 322-331, (2000)
[8] Czechoslovak State Code ČSN 01 0101 - Terms in Quality Control
[9] Carper, K. L. (editor): Forensic Engineering, Elsevier, (1988)
[10] Blockley, D. I. (editor): Engineering Safety, McGraw-Hill Book Europe, (1992)
[11] Drdácký, M.: Optimization of Civil Engineering Structures, (in Czech), Proc. of the IASS Seminar Prague, pp. 44-51, February (1987)

Advanced method to control the safety factor of a civil structure after a collision

Ane de Boer

Civil Engineering Division, Ministry of Transport, Public Works and Water Management, P.O. Box 20.000, NL-3502 LA UTRECHT, The Netherlands, Tel +31 30 285 7699, Fax +31 30 2888 419, Email: a.dboer@bwd.rws.minvenw.nl

SUMMARY

Collisions in civil structures are till today not so common and of course the hope is that it will not become common. However, these collisions can happen in an internal or in an external way. The negative effect of collisions to structures in the road or railway infrastructure of a country can be very large. Especially in a country like the Netherlands with a high density of people with a lot of infrastructures. Missing a structure in a important road or railway link can be a great problem for the traffic stream every day again.

A second order effect by collisions is the impact on the safety of a structure. When the structure has been demolished, it will be reconstruct, when the structure is still needed. Mostly a part of the structure will be demolished and not in total. So the structure will partly be used. An example is for instance a block on two lanes of a six lane fly-over. At that moment an reanalyses has to be made to get an idea of how many lanes have to be blocked with a sufficient reliability for the rest of the lanes.

In the design stage of a structure, the common way to get a safety factor for a structure is to check the results of an analysis with the current checking code. This can be the EURO code or a National code. These checking codes can be changed during the lifetime of the structure. However, after a lifetime of some years the structure can be changed too. It is also very common that some structure parts have been added. A digital form of geometry data of the structure is not always available. To get the as-built configuration of the structure drawings can be checked in relation to the current situation, but is very time consuming. The elapse time of this process is large, so looking to some alternatives should reduce the total elapse time.

Scanning technology is a rather new product in the field of civil engineering. However, this technology which is coming originally from the land surveying and the petrol industry is a very handsome tool to get a 3D model of an 'as built' structure. The civil engineering division in the Netherlands maintains a lot of structure. A large part of these structures are constructed in the pre-digital period. So, the CAE models of these structures are not kept on a disk or a CD in a digital way as a backup.

The paper describes the scansytem itself, the process of scanning, transforming the clouds of points to surfaces and from surfaces to a basic geometry. Additional to this paper items, some general possibilities of the new technology, dedicated to a civil structure (a steel structures) in detail will be described.

Forensic engineering: the investigation of failures. Thomas Telford, London, 2001.

1. INTRODUCTION

Today the common process in the collision situation is, to get as soon as possible a visualisation model or a FE-model from a backup medium. However structures from the pre-digital period don't have such a backup. It is also very common that the backup system fails because of the change of operational systems of the current computers. The same effects occur when the engineer thinks he can handle this process with the current drawings. The drawings should be up-to-date, otherwise you are loosing information about the structure. Often there are made changes in the structure, not only during construction of the structure but also afterwards.

Till today the most effective procedure is a combination of hand made measurements and the available drawings. An alternative procedure on this hand made procedure is to take photographs or using photogrammetry. By coupling photographs or photogrammetry pictures a realistic image can be made of the structure. However the above mentioned procedure isn't very accurate in case of large structures. The net range of each picture is about 10 meters. With span lengths of 200-300 meter it takes a lot of pictures with a high risk of accuracy loss. Beside the accuracy problem there is still a time problem, coupling the pictures takes a lot of time, so minimising the elapse time can be solved partly.

In case of the collision of a structure, the structure will be damaged locally as well as globally. Global deformations can be seen on this type of pictures, but local deformations are often skipped especially in the hand measurement situation. An alternative procedure should be found and is part of this contribution.

2. PETROCHEMICAL INDUSTRY

In the petrochemical industry the scan technology is used to get an overview of the situation on the petrochemical plants. Of course there are drawings of the plants, but mostly there has been some changes.

The plant manager is not always sure that the situation of the plant has been added to the drawings of the plant. So this scan technology is a tool to get a 3-dimensional overview of the 'as built' situation. During scanning on the site the operator of the scan system gets already an impression of the scan quality.

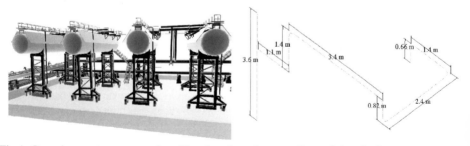

Fig 1. Overview part storage tanks with a drawing of system lines of the pipelines

Back to the office the operator and its assistants can assemble the scans to an overall scan. Now the overall scan can be transferred to, for instance a visualisation package. At this time the end user gets already a good impression of the situation. Transferring these data to a

FE package delivers the basic geometry of a FE model, ready with additional aspects like supports, material properties and loading cases for a simple FE analyses.

The above mentioned procedure has become interested for the Civil Engineering Division in the Netherlands of the Ministry of Public Works. The last years there were some accidents in the civil structure corner, which took a lot of time to get a FE model of the structure. The quality of the FE model should be high. Measurements in steel sections shouldn't have larger tolerances then some mm, otherwise the method wouldn't be successful. So in this way the engineer hasn't to start from scratch with the 2D drawings. The selling company has promised that the scanner was very accurate, so an orientation phase was set-up to control the mentioned scanitems on civil structures. For this reason a Dutch projectteam visited the scan company Cyra in California, US to scan a part of a steelbridge in Oakland. On behalf of the good results of the Oakland bridge, the second part of the pilot, to scan a total bridge in the Netherlands became realistic. After this orientation phase a steel bridge with some collision damage has been scanned. The process of this last bridge is the main part of this contribution.

3. ADVANCED SCAN SYTEM

The scan system consists of a portable auto-scanning laser (LIDAR) and PC software[1]. This combination seems to make laser scanning a good alternative for acquiring accurate measurements and producing 3D models. The CyraX system can measure, visualise and model large structures and sites with an unprecedented combination of speed, completeness and accuracy. First of all a video shot sends an image to the laptop PC. The user selects the desired area of scanning and measurement density of the scan. Depending the density, a scan can be made within just a few minutes and during the scanning the user can already look on the PC if the scan results into a good form of 3D points. This 3D point clouds are a valuable deliverable which represents a 3D virtual model of an existing site or structure. This models contains already 3D geometry information and point to point distances. From this 3D virtual model 2D plane drawings and wire meshes van be extracted. The integrated software which drives the CyraX system will also combine multiple scans to get an total overall scan. Registration of these scans can be done on different ways, for instance the pair-wise assembling method or the global point system coupling method. The 3D point of clouds can be converted semi-automatically into geometrical shapes like: cylinders, spheres, surfaces and steel profiles. The generated models can be exported as 3D models and 2D drawings directly into AutoCAD® and MicroStation® and other CAD or rendering software.
An overview of the system will be given in figure 2.

Fig. 2 Scansystem overview

Beside the scansystem, a tripod, a power unit, when there is no electricity available and the already mentioned laptop PC has been shown.

The scansystem offers many advantages, which are already proved in the petrochemical industry. Traditionally sites or structures will be measured by hands, surveying and photogrammetric methods of measuring.

This scan method promises in relation to the other methods:

- accurate geometry
- saving elapse time
- output for different options
- improved safety
- cost reduction

The range of each scan has a maximum length of 100m, a angle of 40 along the vertical and horizontal alignment. Each scan contains a maximum of 1 million points.

4. ACCURACY TESTS

Of course there has been already done a lot of tests in the petrochemical industry area. Measurements of plants and pipelines are very common and the as-built 3D drawings has been very useful. For the civil engineering industry typical parts like reinforcement bars in a concrete structure, which isn't casted yet or nails in a steel structure could be very important. So the projectteam was focusing on that kind of accuracy tests. Coupling of scans and the accuracy effect on coupling in a global sense of structures with a long span was a second item of interest.

During the orientation phase a start has been made of the first accuracy test. It belongs to the reinforcement of a concrete structure. On top of a pylon with a distance of about 60 meter from the scanner, it was only difficult to reach this part of a structure but also with blocking the underlying road lane. This last item was unacceptable for the road authorities, so the scan system was used.

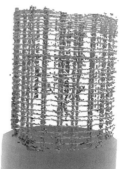

Fig 3. Overview of a fly-over pylon and the reinforcement bar scan

With this scan the diameters (8 and 24 mm) of the reinforcement bars can be measured. Even the amount of the bars can be counted. The supervisors during construction of the structure were very pleased to have these information. The simple clouds of points can be transferred very easy by software to realistic physical properties, like figure 3 shows.

Fig. 4 Close-up of the detail of a steel bridge

A second civil engineering detail is a connection of some steel girders, coupled by bolds, nails or rivets. On the scan we see some flanges, two webs and a lot of rivets, connecting the web and the flanges to each other. Typical aspect of the scan system is the shadow working of parts of the structure, so the back area of a rivet hasn't some points after rotation of the image for information by scanning from one position. These white areas back of each rivet doesn't contain points in this case, only the front area gives some information. For the accuracy test was this aspect important, in real life we only can see the front surface, the back surfaces are still a question.

Focusing to the clouds of points of the rivets it is possible to get a very accurate idea of the rivet to rivet distance in the structure. By zooming in an area with three rivets, the belonging points are cut out from the original scan image. The particular points belonging to each rivethead are grouped and will be transferred to a sphere. The distance between the centres of the axes of each sphere is the rivet to rivet distance in the structure. Figure 5 shows this process in detail.

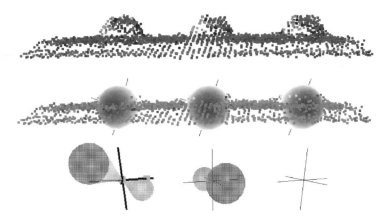

Fig 5. Process of transferring three rivets clouds of points to the sphere axes

In the near future, the software is configured in a way that the software processes automatically a shape proposal for the selected clouds of points. Nevertheless figure 5 shows us first the transferring of the clouds of points to the spheres and after that operation the lower level picture in figure 5 shows only the centre point axes of the spheres. With the measurement option it is now possible to measure the distances between those centre points exactly. The results will be summarised in the following table 1.

Object	Sphere name	Diameter [inch]	Distance [inch]	
Rivet left	Sphere 1	1.52	left-mid	2.98
Rivet middle	Sphere 2	1.55	mid-right	3.02
Rivet right	Sphere 3	1.54	left-right	6.01

Table 1 Measurements spheres

From table 1 it is noticed that the measurements are in inches and not in SI-units, which should be very irregular in the Netherlands. However, this accuracy test has been done in the USA during the first stage of the project, so this structure was situated in Oakland, California. It's assumed that all measurements were made in inches, feet and yards. Looking more close to the results, we see that the diameter of the rivets is about the 1.53 inch and the centre-to-centre distance between the rivets amounts to 3 inches. So the conclusion of this part is rather clear, the accuracy of the clouds of points can configure a well done shape.

The last accuracy test was related to the assembling part of the different scans. From each scan system position some scan images are made each with a certain overlap area from image to image. Back in the office you have to couple these images together to a overall scan. With the use of registration points (objects) the coupling of scans can be made. So there is no problem anymore to scan in civil engineering structures which are mostly larger then only one scan can cover.

These registration points can be part of the structure itself, but can also be more typical scan-points selves, coming from the neighbourhood of the structure, for instance a house or a lamp post. Some characteristic dimensions of the arch of the steel bridge are which is scanned to this accuracy item: length 300.0 m, width 35.0 m and a height of 63.0 meter. From this structure there has been taken 26 scans in 3 different sessions. The first scan session was entering the structure from the North, a second scan session was coming from the South and the last scan session was a scan session from the bottom of the bridge. To get an idea of the total overall scan, see figure 6.

Fig 6. Overview arches of a steel bridge

Coming to the results of the uncertainties of the assembling of the scan files, the horizontal alignment differs in longitudinal direction only 2 degree over a distance of 300 m. The vertical alignment has the same uncertainty, while the assembling of the arch with the lower side of the bridge gives an uncertainty of 1 degree. Weather conditions in march 2000 were very rough, rainy and a lot of wind introducing some mistakes, like missing one scan. This lack of a scan we see back in the middle of the arch of figure 8 on the top of the arch. Nevertheless the scan session was a success and opens a new technology in the civil engineering to develop the geometric part of a model in a quick and accurate way. To look to the additional features like visualisation and the geometric modelling of a structure, a part of the last structure was worked out in that way.

The accuracy items were very successful, so there was no need to add some other accuracy tests in this phase of the orientation of the project.

6. RESULT OF THE SCANNED IMAGES

Only getting a lot of point of clouds is not enough for the civil engineer. After coupling the different images, the overall scan can be used as a visualisation 3D image. By transferring the clouds of points to common shapes a better visualisation image can be found. Today the database of the software knows a lot of pre-defined shapes, so 80 % can be transferred automatically. This ratio can be compared with the translation of CAD data to FE data on the geometric level. Transferring the shape data to the FE data gives the civil engineer a really good base for the geometry of the structure. For this aim a part of the arch of the steel bridge in figure 6 has been transferred and can be shown in figure 7.

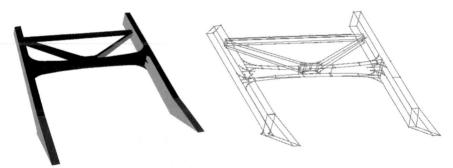

Fig. 7 Fine-tuning visualisation and the geometry of a bridge section

With this figure the connection between the scan software and the FE packages is illustrated. Inside the pre-processor of the FE package the data can be added with the usual FE options like supports, loads, physical and material properties, before a real calculation can be started.

7. AN DEFORMED MAIN GIRDER AFTER A COLLISION

An accident with a container-ship coming from the Rotterdam harbour against a main girder of a steel bridge was an opportunity to get an additional idea of the scanning technology possibilities in civil engineering. The aim of this short scan project was to get in a very fast way the geometry of the deformed structure part, the main girder in this case and to get ready for an FE analysis. The picture in figure 8 gives an idea of the deformations of the main girder.

Fig. 8 Photograph of the deformation of the main girder

The photograph shows very clear a buckling line on the web of the main girder under the level of cross beams. There were already made some handmade measurements, on the positions of the cross beams and between the cross beams along the web in vertical direction, so the results of the scan process can be compared with other results. In this case the handmade measurements were sufficient for the design office of the civil engineering division, to get an repair idea by stretching the girder with a temperature load afterwards. A more exact geometry was not needed for the stretching temperature load. Preparing the scanning process of this main girder, it should be sufficient to scan only the main girder and get some specific cross-sections out of the scans. In this way the geometry part of the main girder could be completely generated in a pre-processor of a FE package. Figure 9 gives an scan impression of the main girder in the orthotropic bridge deck and of course again the deformation of the main girder. This impression is just from one scan position but can be used very well for further investigations and comparing the results with the handmade measurements. It's not necessary to couple a lot of scans because of the local deformation of the main girder. So a position on the north side and one of the south side of the bridge is probably sufficient to get an overall idea of the deformation of the main girder.

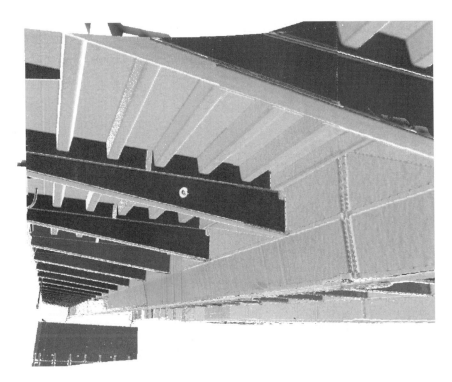

Fig 9 Scan image north side bridgedeck

On the scan image of the north side a very good contrast can be seen between the deformed web. Also the cross beams and the throughs are quite clear. Even the connection by folds of the construction parts of the main girder are clear, so it isn't so difficult to register the deformation related to the amount of cross beams in this case. A relation between the cross beam positions on the photograph and the scan image is comparing.

We know already that the main deformation took place in the area of cross beam 9 till 11, counted from the north side, so in this case we can focus on this area. Zooming in on this area and transferring this area to the global deformation shapes gives the following picture. The fine tuning visualisation part is done by splitting up the area between cross beam 9 and 11 into patches. The point tolerances are minimised of each patch during transferring the data to flat shapes, which is showing in figure 10. The total area counts 7 patches. This means that some discontinuities will occur along the length of the area. This operation causes some difficulties after transferring the geometric data to the FE package. Smoothing the geometric patches to each other in a way that the discontinuities disappear will speed up the modelling time afterwards within the FE package.

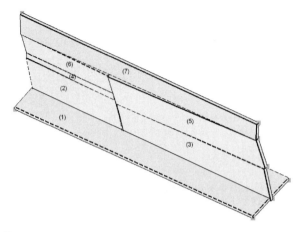

Fig. 10   Shape patches of the deformed area

Comparing the results of the different measurement methods, there can be concluded that the scan system has already very accurate geometrical points. Therefor the patches have also a high accuracy. The maximum deformation was about 35 cm perpendicular to the web, which comes from both measurement methods. The patches of the scan method shows a smoother picture of the deformation of the web and the lower flange. Because of the digital way of scanning and the minimum time consuming method related to the scan method, this method is a preferable alternative method in the future for these kind of accidents.

8. CONCLUSIONS

Accuracy has been improved and in the near future the reach of the scan system will enlarged above the 100m, which is the accurate reach today. At this moment the reach today is sufficient for civil structures. Beside the accuracy, the assembling of the scans is an easy task using the scan software. The overlap from scan to scan will be very important to get the realistic weight factors to the registration points of the assembling scans. The new GPS system will be maximise the accuracy of the results above the pair-wise assembling.

9. ACKNOWLEDGEMENTS

First of all I have to thank the members of Delfttech b.v. in the Netherlands. With the members of this company, I made a start to set-up this pilot project. Last but not least I like to thank the members of the maintenance group near the Suurhoff bridge in the Netherlands for their help during scanning the bridge in the Netherlands.

10. REFERENCES

1.   Cyrax manual, Getting Started/Quick Reference, Model CGP 2.1, Cyra Technologies, Inc., 2000

2.   Cyra manual, Tutorials, Model CGP 2.1, Cyra Technologies, Inc., 2000

Collapse: The erosion of factors of safety to 0.999

JONATHAN G M WOOD, Director, Structural Studies & Design,
RAEng Visiting Professor, Aston University. jonathan@ss-design.demon.co.uk

INTRODUCTION

Forensic Engineering arises from the investigation of structural problems after a failure, when the lawyers get involved. As Engineers we need to apply the lessons from our forensic work to reduce the risks of failures occurring unexpectedly, with substantial economic and social costs.

The premature deterioration of concrete structures has become a costly problem. Government and research in the UK is increasingly focused on sustainability. This will require improved durability design [1] for new construction, as distinct from the "build - scrap - recycle" approach of manufacturing industry.

A more difficult engineering challenge is the safe, cost effective maintenance of our existing deteriorating and substandard built infrastructure with minimum disruption. The Standing Committee on Structural Safety (SCOSS) [2] has drawn attention to the risks. The hazards have been highlighted by a number of collapses in the UK [3] [4] and overseas [5]. In their 13th Report SCOSS have highlighted the limitations of design codes, the way they can be interpreted and the increased risks of failure which can arise with innovative construction outside the scope of standards. The approach to the control of risk which they recommend in design and construction, also needs to be applied when appraising old structures and embarking on remedial work.

This paper focuses on the assessment of falling reserves of strength in deteriorating concrete structures, with examples related to flat slab punching shear, Figure 1.

Figure 1. Punching Shear Failure Developed into Progressive Collapse.

Forensic engineering: the investigation of failures. Thomas Telford, London, 2001.

Failures occur when the initial factor of safety intended by the designer and code committee becomes eroded in just one structural element to 0.999. This need not be a matter of concern if the failure mode is identifiable, ductile and redistributes to other robust load paths. Collapse occurs when this redistribution creates a mechanism. When the mechanism develops into progressive collapse the results can be catastrophic, as when 500 were killed at the Sampoong department store [6] following an initial punching shear failure.

Codified design, as in BS8110 [7], is based on simplified:

- characteristic loads, multiplied by load factors γ_f,
- structural analysis, assuming ductility accommodates variation from actual behaviour,
- characteristic strength of each element divided by a materials factor γ_m.

These factors, based on the old assumption that concrete was 'maintenance free' and that concrete strength increased with time, have no allowance for deterioration of reinforced concrete structural strength with time.

To manage deteriorating structures we need to understand:

- at what level of risk do we say 'it is unsafe'.
- how factors of safety get eroded?
- how we can identify the magnitude of loss

HOW UNSAFE CAN WE TOLERATE?

Safety is not an 'on off' condition, as perceived by the innumerate media and most of the political and administrative establishment. It is a quantitative level of risk, best explained comparatively on a scale related to the chances of being killed on the road (1 in 16,000 a year), by lightning (in 15,000,000 a year) or a meteorite. This approach has been well set out by HSE [8]. In a important new initiative Rochester [9] is setting up an informal Institution of Structural Engineer study group on risk and reliability. This should help develop our approach to risk in structure and the way in which we communicate the risks to clients and the public. We must appreciat that the public expectation that buildings should be "as safe as houses" means that we must air to keep risks very low.

The as-designed level of risk in new construction is set by the simplified rules in the codes an the partial factors on strength and loading. With conventional design and in-situ construction o well understood concrete structures with regular configuration, detailed to ensure ductility, thi gives a reasonably consistent and acceptably low level of risk. The factors of safety γ_f and γ were crudely calibrated in about 1970 for CP110 to give broadly similar designs to the old CP11 working stress designs. They have only been marginally refined since. The more detaile approach to rationalising factors of safety proposed in CIRIA 63 [10] was not adopted fc BS8110.

BS5400 for bridges has more comprehensively adopted the CIRIA 63 approach, driven in part b the need to adopt a more rational approach in the strengthening of box girders after the Milfor Haven (Cleddau) and Yarra (West Gate) bridge collapses. The Highways Agency has been in th forefront of further development of this reliability based approach [11] to bridge management. The fundamental difficulties in the application of this to concrete structures are in evaluating the initial strength condition with design and construction errors and predicting the rate and magnitude of strength loss as deterioration develops.

The drafting of Eurocodes has renewed the debate on appropriate factors of safety. The principle of reliability based partial factors is agreed.   The appropriate values to achieve consistent reliability with different materials and structural forms remains a keenly debated, but perhaps inadequately researched, issue.   There are strong pressures for economy and simplicity to enhance the competitive advantages of different materials.

The reliability targets of current codes for ductile in-situ concrete structures provides a point of reference against which to judge more complex structures in our ageing infrastructure.

Structures may not be 'as good as new', so what reduction in factors of safety is acceptable? 25% overstress was the traditional yardstick for the limit for acceptance, after a detailed inspection and appraisal to refine loading, analysis and strength.   However the market pressures to slim down designs have already led to reductions in code partial factors and it may be argued that the codes are now being set at a level where no further reduction in factors is acceptable in appraisal. It is essential that those who calibrate the partial factors in codes make clear:

-      the limits of the range of structural configurations for which they are valid,
-      the level of risk they adopt in setting them,
-      the increase in risk when factors are reduced.

APPRAISING THE EROSION OF BS8110 FACTORS OF SAFETY

IStructE 'Appraisal of existing structures' [12] provides good guidance on the principles to be followed, but more advice is needed when structural forms differ from those in the codes and when deterioration changes stress distributions, the effectiveness of detailing or reduces bond strength.

While BS 8110 is the starting point for determining the best estimate of current strength for appraisal, it does not provide a consistent yardstick for structures at the margins of or outside the range of structural forms considered by the code committee. Many older structures have features which do not meet detailing requirements and some have design faults. There have been major changes in design requirements particularly in shear, as research has uncovered deficiencies in old design rules, eg Regan [13, 14]. All involved in appraisal must be aware of these changes, which can substantially reduce the factors of safety to below the reference level required for current design.   Engineers have a responsibility to make their clients aware of these reduced factors, despite some clients preference for commissioning the checking of old structures only to the codes to which they were designed.

What does γ_{ft}, the partial factor on load, cover?

It is surprising that, despite the growing sophistication of loading calculations (eg wind and snow load), structural analysis and strength calculations, the broad brush partial factors in BS8110 remain unrefined and applied regardless of structural form and construction quality.

The presumption in appraisal is that these factors can be reduced as analysis is refined, but in some circumstances it is more appropriate to increase them if consistency of risks is to be achieved.

In BS8110 2.4.1.3 it states that the γ_f the partial factor on load is to *"take account of:*

-        *unconsidered possible increases in load,*
-        *inaccurate assessments of load effects,*
-        *unforeseen stress redistribution,*
-        *variations in dimensional accuracy, and*
-        *the importance of the limit state being considered".*

BS8110 uses $\gamma_f = 1.4$ for dead loads and $\gamma_f = 1.6$ for imposed live loads for load combination 1, with reductions in combination with wind. Temperature effects, misfit or foundation differential settlement are not explicitly considered. Clause 2.4.1.4 refers to loads during construction, but not to the effects of tolerance and misfits during construction, which many would expect to be covered by γ_f as ' *inaccurate assessments of load effects'* or *'variations in dimensional accuracy'*. For lift slab punching shear the construction tolerances on the wedges connecting columns to the slab can substantially erode the γ_f, leaving only 1.1 to 1.2 for other factors. Some structures are much more sensitive than others to overall and differential temperatures and to differential foundation settlements. Because of BS8110's failure to consider them explicitly they can erode factors of safety.

In most normal in-situ construction the detailing ensures that when technical 'failure' occurs from excessive strains in the concrete or steel the load carrying capacity is largely retained, Figure 2 Large deflections and cracking can redistribute surplus moments and shears to other load paths. For ductile structures with a range of load paths, errors in the assumed load distribution are not important and do not influence the risk of collapse as distinct from technical 'failure'. For these structures little of γ_f is needed to cover *"inaccurate assessments of load effects, and unforeseen stress redistribution"*. The limit state is of visible distress and excessive deflection, not collapse so γ_f contains little margin for structures with limit states of more *"importance"*.

Figure 2. Ductile and Brittle Failure

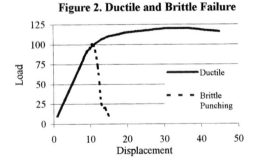

Punching shear failures in flat slabs are sudden brittle and shed all load to the adjacent elements. Any underestimate in analysis of reactions and transfer of moments from slab to columns erodes the margins against a brittle shear failure, which can precipitate progressive collapse, the most *"important"* of limit states. BS8110 permits a wide range of analytical methods. The redistribution of reactions and moments from interior columns makes errors in the assessment of upper bound load effects probable.

Clearly the BS8110 use of uniform γ_f irrespective of structural characteristics results in very wide variation in the risks of collapse with different structural types. In appraisal it is essential that the sensitivities of structures are identified and load effects of greater magnitude than for a reference in-situ ductile structure are quantified in structural appraisals to achieve consistent risk.

What does γ_m, the partial factor on materials, cover?

The materials factor γ_m in BS8110: 1985 2.4.2.2 takes *"account of :*

- *differences between actual and laboratory values, local weaknesses,*
- *and inaccuracies in assessment of the resistance of sections.*
- *It also takes into account the limit state being con sidered."*

Curiously in BS8110: Table 2.2 γ_m is:

1.5 for concrete in flexure and axial load,

but is only   1.25 for shear strength without shear reinforcement, a mode of failure which is sudden and brittle.

Strength is calculated on the basis of nominal dimensions. Tolerances on thickness and cover are defined in standards. Slabs, constructed within tolerances at the lower bound of thickness with upper bound of depth of cover, can be 10 to 20% under strength, in flexure and shear. Beams are deep enough to have a low sensitivity to this, but the thinner the slab the greater the sensitivity and risks because of the uniform γ_m.

While Section 2.3 of BS8110 requires rigorous inspection standards, much construction practice had lower standards of quality control. Surveys of reinforcement cover in structures show that the actual variation is far wider than the specified tolerances in standards leading to a greater risk which must be quantified in appraisal. Covermeter surveys can be used to establish actual cover range. The potential effects can be determined from strength checks at characteristic and likely extreme values of cover to determine the sensitivity of strength to as-built variations.

Determining the actual strength of concrete often identifies additional reserves of strength in a structure. However one needs to ensure that the variability of construction did not lead to patches of lower strength. Sufficient cores must be taken to establish the pattern of variation in strength. A wide scatter will indicate the risk of lower strength areas and a need for more extensive coring than where good quality control is indicated. Low strength concrete is more susceptible to deterioration and so coring into areas were signs of premature spalling or frost damage are developing can target sampling on weak areas in damp structures. Non destructive techniques like Schmidt hammer and Pundit pulse echo can also pick up areas of relatively low strength for coring.

EFFECTS OF DETERIORATION ON STRENGTH
Once deterioration starts there is a risk of factors of safety being eroded. However much deterioration is in low stressed areas, is cosmetic or is leading to serviceability problems and risks from spalling, without the overall strength being reduced. It is only by relating the patterns of deterioration in a structure to the particular sensitivity of the reinforcement detailing that the consequences of deterioration and influence on safety can be judged.

The IStructE guidance 'Structural effects of ASR' [15] sets out an approach for identifying the severity of deterioration, the sensitivity of structural detailing to deterioration and the potential consequences. This enables an appropriate level of structure specific structural evaluation, inspection and monitoring to be carried out depending on a rating from 'mild' to 'very severe'. The same approach works well with other forms of deterioration from corrosion, frost etc. However structures in the 'severe' range need specialist investigation related to fundamental research and case study analysis [16] rather than the simplified guidance.

A key element in the ASR approach is that the location, methods and extent of investigation are based on a prior structural evaluation. The great advantage of ASR is that the progressively developing cracking, from the interaction of variable expansions with structural strains, shows on the surface to make it inspectable and monitorable.

Corrosion, particularly under surfacing, is more difficult to identify in superficial inspection. On many structures 'corrosion only' surveys are carried out by specialists without a clear structure specific brief from a structural engineer. These often fail to check for other forms of deterioration and defects of construction and do not focus on the areas most structurally sensitive to corrosion. Inappropriate inspection increases risk by delaying proper assessment.

The most damaging structural effect of corrosion, frost and ASR is delamination of cover leading to loss of bond. This can seriously weaken a structure, eg where a car park or bridge relies on the anchorage of top steel around a punching shear perimeter [17]. Repair of this delamination is particularly difficult as repairs introduce a plane of weakness below the top steel. Both the deterioration and the cutting out for repair can redistribute loads and create stress concentrations which are of particular concern with brittle elements in a structure.

The loss of area from reinforcement corrosion and the embrittlement from pitting can become critical. The development of corrosion at transverse cracks over supports or in half joints and the loss of shear steel anchorage from corrosion on corner bends need particular attention.

In reviewing the safety of these situations the reliability of inspection and monitoring of deterioration needs to be specifically considered. Chlorides seeping below surfacing can seriously weaken a structure before superficial inspection detects it. Any breakdown of surfacing in such an area needs thorough structural investigation of the condition of steel and bond, not just superficial local corrosion check and patch repair.

The major difficulty in the management of structures is the lack of detailed guidance on the effects of deterioration and repair on strength, ductility and modes of failure. The sparsity of research to date makes the development of advice, other than generalised expressions of principle, difficult.

PROGRESSIVE COLLAPSE

Progressive collapse was seldom explicitly considered before Ronan Point. The robustness requirements in CP110 and now Clause 2.2.2.2 of BS8110, which *"may generally prevent accidents"*, were not considered in many early structures. When old structures do not meet this requirement they should be required to have higher factors when appraised or should be subject to the specific evaluation of the hazards of progressive collapse, as advocated by SCOSS for design.

When a flat slab without shear steel fails in punching the load is transferred to surrounding supports which then tend to fail. Ties to BS8110 tend to ensure that this spread continues. In design, this risk of progressive collapse can be simply prevented and effective tying of the structure achieved, by providing a sufficient area of bottom steel directly over the column [18]. This transforms the brittle failure to a ductile mode, with the majority of the reaction supported after the initial punching. The adoption of this for all new design with would remove the anomaly of a permitted brittle detail in BS8110 and the disproportionate hazards inherent in current designs. For existing structures with the brittle detail and for which specific evaluation indicates a risk of progressive collapse, the maintenance of full factors of safety is prudent in appraisal.

CONCLUSIONS

The adoption of uniform simplified partial factors in BS8110 has resulted in a wide range of risk of failure and collapse in structures as designed. This is increased by variability in construction.

The use of superseded codes or the 'generally applicable' clauses and simplifications in BS8110 for unusually proportioned or innovative structures and for obsolete designs can lead to substantial errors in determining risks in appraisal.

When structural elements are not ductile, their sensitivity to temperature, foundation settlement and misfit in construction must be evaluated. The upper bounds of load effects and stress concentration must be established in analysis. This also applies where deterioration may embrittle a ductile element.

Progressive collapse hazards needs to be checked explicitly for the structure as-built and with consideration of likely patterns of deterioration. The BS8110 robustness requirements should be considered as a possible, but not certain, means of preventing it.

Punching shear in flat slabs of all types needs rigorous evaluation because of their brittleness, sensitivity to deterioration and tendency to progressive collapse.

REFERENCES

1   Wood J G M, "Durability design: applying data from materials research and deteriorated structures", Bridge Management 3, Harding J, Spon, London, April 1996.
2   SCOSS, "Standing Committee on Structural Safety Reports", 11th 1997, 12th 1999 and 13th 2001.

3   Kellerman J, "Pipers Row Car Park", Concrete Car Parks Conference, BCA, Sept.1997
4   Woodward R J and Williams F W, "Collapse of Ynys-Y-Gwas Bridge, West Glamorgan".
        pp 635 - 669, Proc. Instn. Civ. Engrs, Pt 1, 4. August 1988.
5   Carper K L, "Current structural safety topics in North America", The Structural Engineer,
        June 1998.
6   Garner N J et al, "What can we learn from the Sampoong Department store collapse",
        International Workshop on Punching Shear Capacity of RC Slabs, Stockholm, 2000.
7   BS8110-1985, "Structural use of concrete", BSI London, 1985.
8   HSE, "Reducing risks, protecting people", HSE Discussion Document, 1999.
9   Rochester T, "New informal study group on risk and reliability", Structural Engineer
        79/15, August 2001.
10   CIRIA 63, "Rationalisation of factors of safety", CIRIA London, 1977.
11   Das P C, "Reliability based bridge management procedures", 'Bridge Management 4',
        Ryall M J, Thomas Telford, 2000.
12   IStructE, 'Appraisal of Existing Structures', Second Edition, SETO, October 1996,
13   Regan P E, "Research on shear: a benefit to humanity or a waste of time?", Structural
        Engineer 71, 19, 5, October 1993.
14   Regan P E, "Behaviour of reinforced concrete flat slabs", CIRIA Report 89, 1981.
15   The Institution of Structural Engineers, "Structural Effects of Alkali-Silica Reaction,
        Technical Guidance on the Appraisal of Existing Structures", London, 1992.
16   Wood J G M & Johnson R A, "The Appraisal and Maintenance of Structures with Alkali
        Silica Reaction." The Structural Engineer 71, 2 pp 19-23, January 1993
17   Chana P S and Desai S B, "Design of shear reinforcement against punching", Structural
        Engineer, 70/9/ 5, May 1992.
18   Mitchell D and Cook W D, "Preventing progressive collapse of slab structures", J of
        Structural Engineering 110/7, ASCE, July 1984.

A knowledge based survey of buildings for structural reliability evaluations

ROBERTO E. GORI, Dipartimento di Costruzioni e Trasporti, Università di Padova, Italy

ABSTRACT

The last disastrous events occurred also recently in Italy, characterised from the collapse of constructions also built in relatively recent, in absence of exceptional actions, have proposed the problem of the dangerousness of great part of the existing building patrimony and the necessity therefore to develop criteria of control of fast type.

As it is well known, in presence of only permanent actions and of those variable ones within the serviceability limits, and excluding therefore causes as the fire, the collision of vehicles or aeroplanes, the earthquake or other natural events of extraordinary entity, have taken place numerous structural failures, also without warning, often arousing a reasonable alarm in the public opinion, to the point to speed up the institution of instruments of law in a position to preventing such catastrophic events. In this paper one proposed methodological of evaluation of the state of structural reliability of a construction is introduced that re-enters in the spirit of performance foretold from such legislative instruments. The proposal is confirmed from one series of useful numerical tests for its validation.

INTRODUCTION – METHODS OF EVALUATION OF THE RELIABILITY OF EXISTING BUILDINGS

The reliability of the existing buildings is a argument of great interest of the present time. The greater part of the existing building patrimony in Italy has been constructed after the second world war and in particular with the building boom around the years ' 70.

During the post-war reconstruction new material as reinforced concrete, without adapted experimental support, has been used. In any case, without taking into consideration the associated cultural or construction aspects that may have conditioned the correct execution of these buildings, it turns out obvious today that we have an increasing number of "obsolete" constructions of more than 40 years. The question rises spontaneous: are such buildings still safe?

The recent landslides happened in Italy, with numerous loss of human life, have confirmed the existence of relative problems of the durability of the constructions. To these constructions the minimal level of performances must be guaranteed, and must be maintained during the entire service life. Since the remarkable resonance of these tragedies, there have been many arguments and debates on avoiding or at least reducing the future landslide probability. The lack of adequate norms that can take part to guarantee the safety of the existing buildings has pushed the Italian government to propose (law design 4339-bis, currently still with to the Italian Congress) "the institution of the issue of the building", an obligatory document for the assessment of the safety conditions. The target is to propose one methodology in order to estimate the emergency of the existing buildings. Such proposal makes reference to various important experiences at international level under the normative and methodological aspects.

Forensic engineering: the investigation of failures. Thomas Telford, London, 2001.

Before describing these experiences, it is well to distinguish the building patrimony in two categories: 1) future constructions or of recent edification, for which there exist codified normative instruments that guarantee levels of adapted qualities and safety through design, executive processes, of control and maintenance; 2) existing constructions or of not recent edification, for which the minimal level of safety must be guaranteed.

The various analysed international experiences comprise normative or methodological, in order to guarantee the safety of existing constructions. From the searches a heterogeneous picture turns out rather, in how much every nation has one own "constructive history". But the importance of the successive maintenance aspect to the construction turns out to be obvious. In countries as France and United Kingdom in which, beyond having a control taken care of the construction process through specific responsibilities, low maintenance programs of the realised work. These maintenance programs and the introduction of new professional figures, indicated from recent directives the EEC 92/57, do not make other that to assert one already consolidated constructive structure based on the design and executive quality. Moreover they are present in both countries, even if in various shape, in which a series of data regarding the characteristics of the piece of real estate, comprised all the interventions and the modifications carried out in the course of the years, let alone indications on the state of conservation. These documents are essential in the processes of real estate transition to guarantee future buying.

Also in Italy with the law on the public works n.109/94, following the European Directives, and the recent general regulations of performance introduced with DPR 554/99, has been introduced the plan of maintenance of the building, in which periodic controls are previewed. Unfortunately such norm is applicable exclusively in the realisation of public works and reference to the importance and the specificity of the intervention. Other interesting proposals come from countries with various truths as in the case of Hong Kong. Here the Buildings Department has supplied to the institution of one outline for the inspection of the safety of the existing buildings having an advanced age to 20 years. This control has become necessary considering the enormous state of degradation in which they pour numerous constructions, between which many of "illicit" and therefore not authorised. Of absolute importance is the adopted process of evaluation. One first called phase general evaluation allows control, through visual inspections, the state of conservation of the construction expressing, through a gravity index, the advised interventions. The second phase instead, called detailed control, comes only executed if demanded from before, and more regards an taken care by inspection with successive structural analysis. This logical process of levels allows them to eliminate the cases with little defects in order to concentrate the attention on the more important and problematic cases.

Similar to the process seen for Hong Kong is the one proposed from the *American Society of Civil Engineers* (ASCE) in which are supplied a methodology for the evaluation of the structural conditions of existing buildings in concrete, metal, masonry and wood, through a preliminary evaluation and if necessary through a detailed evaluation. Other specific indications come then supplied from predisposed international norms from International Organisation for Standardisation (ISO), as the ISO/FDIS 2394 and ISO/CD 13822 (rev), in still not definitive writing, in which the criteria come established for the evaluation of the existing structures and serve as a guide for possible national norms. As evaluation methodologies, even if relative to various application fields, they have been shortly described the experiences had from Manfred Wicke (University of Innsbruck) and the Italian GNDT (National Group for the Defence from Earthquakes) of the CNR. The first one has processed a method in order to estimate the state of conservation of the street bridges in reinforced concrete while the second has predisposed a method for the survey of the exposure and seismic vulnerability of the buildings. It does not go then forgotten the issue of the building realised from Rome Administration for the assessment of the static-functional, obligatory consistency for all the buildings and restored every eight years. In such issue a series of indications goes brought back all that go from the identification to the consistency of the structures carrying until the writing of a synthetic relation in which the technical person in charge to the writing of the issue it proposes to place or less the building under observation. All these norms and methods of evaluation are served from reference for the formulation of the proposed methodology. This proposal has the target create a method of surveying of "speedy" type through which it can be reached to one reliable evaluation on the state of conservation of the construction and debit of the first indications on the interventions to follow.

The process of evaluation, carried out for successive levels, previews initially a preliminary evaluation in order to establish a first approach with the object constructed through the acquisition of a series of identification data and of data obtained from mostly visual inspections or with I use it of simple instruments.

The evaluation method in this phase can be considered of quantitative/qualitative type. It previews a classification of the structural typologies, depending on the constructive system and the used materials, reinforced concrete, steel and wood. To each typology it has been attributed a coefficient, defined as "weight" of the structure (Ps), considering the typology in reference to the carrying capacity of the entire structure. The true relief follows, just characterising, for every present structural typology in the construction, the intensity and the extension of the found defects.

The defects are reported to lesions, cracks, gaps, deformations, corrosions, etc., depending on the analysed structural typology. Also in this case they have been assigned of the coefficients weighed to the level of intensity (Pd) and extension (Ps) of the defect. During the inspecting visit incidental changes of use or structural modifications not authorised or not previewed in the original plan must be verified also, taken into account through corrective coefficients defined as value of use (Vu) and value of modification (Vm). All this series of weighed parameters serve in order to express the Total Index of Structural Gravity (It) that it represents the maximum value between all the Partial Indices of Structural Gravity (Ip) founded. The result of the evaluation will be determined from the value caught up from the It index, to which it will correspond one of the six previewed classes of damage. To every class it has been assigned a synthetic judgement that expresses the level of caught up damage and gives the first indications on the advised interventions.

For having then a general picture on the state of consistency of the building evaluation object it is necessary to fill a document, called "Issue of the Building", in which are collected the necessary data for being able to express a judgement of final merit with the indication of the possible successive interventions to take. Moreover such issue will serve as base for future restorations or retrofitting in relation to the interventions brought to the building in the course of the years. Only if demanded from the recommendations of the preliminary evaluation, detailed evaluation have to be executed. This evaluation has the scope to assess the reasons of the endured damage and to establish the current level of safety. To make that, it will be necessary to execute detailed inspections in order to determine the loads, the properties and the strengths of the materials. Structural analyses will be then executed, by means of quasi-probabilistic calculations to the limit states. The results of all the verifications carried out in the detailed evaluation will be attached to the issue of the building.

A METHODOLOGICAL PROPOSAL FOR EVALUATION PROCEDURES

In spite of the variety of the normative proposals and the evaluation methods seen previously it can evidence a not homogenous interpretation of structural safety in the existing buildings, characterised by the common aspects that will serve as base for a methodological proposal of evaluation. An emerged aspect of absolute importance indifferently is that one of the prevention, through plans of maintenance of the work, previewed from the European norm and applied nearly uniform in all the states members. Such aspect but enters in single function in the case of construction of new buildings or the restructure of existing buildings. Another aspect of common importance is the importance of the role that must assume the customer, leading or owner, as knows the characteristics and the limits of the building in which it is found to living, carrying out periodic inspections of simple execution. It is obvious that as far as the buildings of recent or new construction, thanks to the introduction of more and more severe norms for the execution and quality control of the materials of the constructions, the maintenance programs, turn out more difficult tragic events for unexpected failures.

The problem rises instead for all those buildings, making part of the enormous existing building patrimony, that for several reasons cannot be considered sure under the structural aspect. For such buildings a control and often the improving interventions for guaranty the reliability during their entire life are necessary. The method proposed for the evaluation of the reliability of the existing buildings takes cue from the studied cases and previews a "multilevel" approach.

The base of the process previews a *Preliminary Evaluation* continuation from a *Detailed Evaluation*. The *Preliminary Evaluation* establishes the first approach with the building through the acquisition of a series of identification data and data obtained from mostly relative visual surveys to the state of conservation of the structures. The evaluation method of qualitative/quantitative type supplies a more general definition of the conditions of the structures establishing necessity and priority for a further detailed analysis, if necessary.

The *Detailed Evaluation* is only executed if it is demanded from the recommendations of the preliminary evaluation; further integrating data are collected, relative to the loading actions, to the tests on the materials, the property of the materials, etc., in order then to execute structural analyses through quasi-probabilistic approaches by models to the limit states, with the successive determination of the level of reliability. Subsequently to the evaluation it is written up a issue containing all the relative data of the inspecting visit: identification data, carried out, turned out controls of the evaluations, judgements, interventions, etc. Once compiled such issue will be guarded inside of the building for possible successive inspections.

The exposed evaluation method is applicable indifferently to whichever constructed building and does not take in consideration the pure normative, relative aspects to the modality, to the times of application and the inspecting cycles, that they will be asked to the legislator.

Preliminary Evaluation

The Preliminary Evaluation is identified as the first approach for the verification of the state of conservation of the existing structures. Beyond to a series of data relative to the identification of the building, the constructive and structural systems and the possible present defect in the structure are defined, through visual surveys observations or with simple instruments. The collected information are relative to the defects of surface, deformations, breaches, lesions, corrosions, etc., depending on the employed structural typology. The result of the preliminary inspection are expressed in terms of qualitative judgement through a classification in six levels that expresses the gravity of the damage of the building. To the classification correspond, consequently, the state of conservation and the first indications on the interventions to take. In order to determine to which class belongs a building, reference is made to the Maximum Index of Structural Gravity (Im), that expresses the higher value between all the Partial Indices of Structural Gravity (Ip) found for every floor. This value takes into account a series of weighed parameters derived from the structural typology, the intensity and the extension of the found defect.

Acquisition of the technical and documentary data
The first data that are acquired are relative to the identification of the piece of real estate. Such data comprise: characteristics of the real estate complex, identification of the manufactured object of assessment second of the historical classifications, presence of not authorised buildings and over-elevations, modifications of static importance, description of near buildings, the geometric characteristics of the building, location, identification data the real estate units, town technical data, presence of certifications (technical plans, fire protection, load tests, etc), data on planners and constructors, availability of technical documents (original plans, varying, relations, etc). All these data go brought back in the appropriate cards previewed for the issue of the building.

First objective verifications
If the collected documentation is complete one can already have a enough precise idea of the type of building that we go to inspect. In the majority of the cases, for not recent constructions, many data will turn out incomplete or lacking. This does not preclude the goodness of the final evaluation even if lacking data will be listed in appropriate cards to enclose to the final considerations like data to acquire in the successive inspections.

Proposed Methods of Evaluation
The proposed method of evaluation takes cue from the cases examined previously. The introduction of a series of coefficients, attributed for structural typology, intensity and extension of the defect, serves for having the most possible objective and comparable result, independently from the taken building type in examination. The structural typologies have been divided into four great categories: structures of elevation, horizontal structures, covers and stairs. It has been inserted a fifth category relative to other elements, where are inserted divisions or masonry walls, cornices, cantilevers or other structural elements of secondary importance. Such elements, even if not determining, are considered important to the aim of the final total evaluation.

The structural typologies vary with the constructive systems and the used materials, as reinforced concrete, steel and wood. To each typology it has been attributed a coefficient, defined as "weight" of the structure (Ps), varying from 1 to 10, that measures the importance of the typology considered in reference to the carrying capacity of the entire structure. During evaluation, this Ps value can be varied or adapted according to the characteristics of the structure under inspection. An example of variation regards the structural typologies with a prevailing character, compared with the entire constructive system, as in the case of large span covers or floors.

Defined the structural typologies, we must pass to the description of the defects. The defects are divided in six levels according to the found intensity and express a first qualitative judgement of the state of the analysed structure and the consequences on the carrying capacity and the durability. For each level it has been attributed a coefficient, that it goes from a minimum of 0.10 until to a maximum of 1.00, in order to define the *weight of the defect* (Pd), like best evidenced in table 1.

Intensity of the defect	Assigned weight of intensity of defect (Pd)
1 – No defect	0,10
2 – Little defect	0,20
3 – Medium defect	0,40
4 – Severe defect	0,60
5 – Very severe defect	0,80
6 - Total loss	1,00

Table 1: Classification of intensity level of the defect

Another aspect that must be taken into account in order to estimate the safety in the structures is relative to the extension or frequency of the defect. *It* is defined in three qualitative levels: little frequent, frequent and very frequent, like shown in table 2. Also in this case to each level it is assigned a coefficient that indicates the *weight of the extension or frequency (Pe)* and measures, for every found defect, the amount in relation to the extension of the entire structural typology under examination. If the considered defect is quantifiable with an area (es. gap or lack) the extension in relationship to the surface of the analysed structure is evaluated; if, on the contrary, the defect is quantifiable numerically or linearly (i.e. cracks) the frequency of manifestation is measured. In indicative way, the following levels of extension in relationship to the considered surface can be considered: little, not frequent (< 10%); frequent (from 10% to 30%); very frequent (> 30%). This element of evaluation but can turn out complicated when on the same surface can be present more defects, which can be evaluated by extension or by frequency. In this case the appropriated methodology could be to measure the extension of the more serious defect. The *Pe* value in the cases "not frequent" and "frequent" is not always constant, but it varies in relation to the levels of the defect. More a defect is serious and therefore evident, more the value of the extension increases. In this way it has given more importance to the serious but little frequent defects. The coefficients are therefore divided in four groups reported respectively to the levels of intensity 1-2-3, 4, 5 and 6.

Frequency of the defect	Assigned weight of frequency or extension (Pe)			
	Defect 1-2-3	defect 4	defect 5	defect 6
A – not frequent	0,45	0,60	0,70	0,75
B – frequent	0,70	0,80	0,85	0,90
C – very frequent	1,00	1,00	1,00	1,00

Table 2: Classification of frequency of the defect

With the aid of survey cards for intensity/extension of the defects and with tables of automatic calculation, that we will see subsequently, it is possible to define with facility the *Maximum Index of Structural Gravity (Im)*, that indicates the higher value between all the Partial Indices of Structural Gravity (*Ip*). The determined *Ip* for each storey of the inspected building is given from the higher value turning out from the product of the weighed coefficients relative to the defects, to the structural typology and the extension or frequency of its defect. That can be described through a simple formulation:

$$I_m = \max\{Ip_i\} \quad (i = 1,.., \text{number of storeys}) \tag{1}$$

$$Ip_i = \max \{Ps_k \cdot Pd_k \cdot Pe_k\} \quad (\text{structure k}) \tag{2}$$

where:

Psk	=	weight of the considered structure
Pdk	=	weight of the defect on the considered structure
Pek	=	weight of the extension or frequency of the defect
Ipi	=	partial index of structural gravity (of the single storey)
Im	=	*Ipi* max = is the maximum value found for the storey

For each structural typology, it is possible to report the specific description of the intensity of the defect found during inspection, with the value of the product of the two factors *Ps* and *Pd*. As we have seen in previous formulations (equations 1 and 2) the Maximum Index of Structural Gravity (*Im*) gives back a value of gravity of the present damage in the analysed construction. In order to inquire more in deepened way on the real conditions of safety of the structure it comes held account also of the aspect modifications that can have involved of the variations in the structural conditions. These variations can be relative to the execution of improper works or to changes of destination of use successive to the construction. Therefore two corrective coefficients are introduced defined *Value of use* (*Vu*) and *Value of modification* (*Vm*). Let we see in particular their function.

When it is executed the inspection in order to define intensity and extension of the defects, it is verified also the current destination of use of the storey or the analysed portion of storey. If the use destination is different from that one originally defined or approved of for that type of structure then it will come applied a coefficient of use (*Pu*). This coefficient is only applied to the surface interested from the change of use. The *Pu* coefficient is then standardised in relationship to the entire surface of the storey. The value of use (*Vu*) relative to the storey will be greater than one if the change of use gets worse the loading conditions, minor of one if the change of use reduces the loading conditions previewed in the original plan.
During the preliminary inspection it is always inquired also the possibility of interventions that can have modified or have got worse the structural behaviour of the construction, without that are available relations that certify such modification. Such modifications are classified through qualitative judgements that express the importance of the found modification. An example for absurdity of this variation could be the removal of a pillar, or the opening of a hole of important dimensions in one carrying masonry wall. The definition of a coefficient of modification (*Pm*) reported to the zone interested from the intervention, will go to influence proportionally, following the rules of normalisation seen previously, the value of modification (*Vm*) of the entire one storey. In the following, in table 3, the importance of the intervention and the *Pm* coefficient assigned are reported.

Importance of the structural modifications	Coefficient of modification (Pm)
A – not found or not surveyed	1,0
B – of modest entity	1,2
C – of medium entity	1,4
D – important	1,6
E – very important	1,8

Table 3 - Definition of the parameters of structural modification

Class	Definition of the level of the damage and the advised interventions	Value of "It"
Class 1	**No damage or insignificant damage** No maintenance operation is demanded. Probably a lot of the found defects existed already at the moment of the construction of the building or are superficial and they do not interest the structure. No detailed evaluation is required.	0,00 ÷ 1,00
Class 2	**Light or little meaningful damage** No immediate maintenance operation is required. The found defects do not modify the safety and the durability of the structures. However the normal maintenance must be executed in order to guarantee the durability and the efficiency in the time of the entire building or portion of it. If the complex of the acquired data turns out sufficiently exhausting in order to recognise the state of conservation in which the building is found it is not necessary to proceed to a further detailed evaluation.	0,90 ÷ 2,00
Class 3	**Medium damage** There is no danger for the uses of the building or portion of it, but an extraordinary maintenance operation is necessary, in order to guarantee the durability and the current use of the construction. If the complex of the acquired data turns out sufficiently exhausting in order to recognise the state of conservation of the building it is not necessary to proceed to a further detailed evaluation.	1,80÷ 4,00
Class 4	**Severe damage** Immediate interventions of restructure and repair are necessary in order to guarantee the durability and the current use of the construction. It is necessary a detailed evaluation in order to determine the causes of the damages and for verifying the reliability. The possibility of uses of the construction must be verified during the execution of the works.	3,60 ÷ 6,00
Class 5	**Very severe damage** Urgent measures are necessary in order to guarantee the safety and immediate interventions of repair in order to guarantee the durability and the current use of the construction. A detailed evaluation is necessary. It is not possible to use the construction during the execution of the jobs.	5,60 ÷ 8,00
Class 6	**Very severe damage or total loss** Seen the state in which it pours the construction, immediate interventions of restoration would not guarantee the durability and the use for which it had been constructed. The only possible intervention is the demolition with the reconstruction, if necessary. The damage is so high that no detailed evaluation is required, consisting that it would render such evaluation ineffective.	> 7,50

Table 4: Classification of damages of a building

The Evaluation

After the recording in appropriate automatic forms of the data corresponding to the inspections carried out by means of the exposed methodology, the calculation of partial indices *Ipi*, *Vui*, and *Vmi* is carried out. The Index Total of Structural Gravity (*It*) will be the maximum value between the products of the partial indices. Therefore the expression will be:

$$It = \max\{Ip_i \cdot Vu_i \cdot Vm_i\} \quad (i = 1, ..., \text{number of floors}) \tag{3}$$

$$Vu_i = Pu \cdot \frac{Ss}{S_i} \tag{4}$$

$$Vm_i = Pm \cdot \frac{Sz}{S_i} \tag{5}$$

where:

Pu = Coefficient of use
Pm = Coefficient of modification
Ss = Surface interested from the change of use
Sz = Zone of influence of the modification
S_i = Surface of the considered storey
Vu_i = Value of use of the considered storey
Vm_i = Value of modification of the considered storey
It = Total Index of Structural Gravity

Of course, if the corrective values *Vu* and *Vm* do not find modification to the original loading conditions, the *It* index is equal to the Maximum Index of Structural Gravity (*Im*) seen in equation 1. With the calculated *It* value it will be given a synthetic judgement on the present damage and the advised interventions. Such value will correspond to one of the six established classes of damage, as indicated in table 4.

In the cases in which the value of *It* corresponds to two distinguished classes of damage, the procedure to adopt is to assign however the higher class and to verify their reliability through a general evaluation that takes into account the totality of the collected data.

The *It* index could express a not representative value of all the structure, but only of a storey or one portion of storey. This has had to the applicability of the evaluation method to various constructive typologies that renders impossible to express a medium value or a total sum of products. But at the same time the result reveals the presence of a damage that could

To the end of the evaluation procedure, the result in a single calculation table can be collected several in order to define a diagram of distribution of the partial index of structural gravity referred to the inside of the construction. This table, divided for storey and structural typology, allows to immediately evidence the positioning of the defects and the found degree of gravity. Beyond to the maximum value, it will be possible to formulate a relative average value corresponding to the single structural typology, so as to evidence the degree of manifestation of the damage inside of the same structural typology.

REFERENCES

U.A. (1997),"A Comprensive Approach to Building Management, Maintenance and Safety Inspection of Private Buildings in Hong Kong", *Surveying*, Vol. 6, Issue 8, October.

CENSIS, *Rapporto 1999 sulla situazione sociale dell'Italia.*

CNR (1993), "Rilevamento dell'esposizione e vulnerabilità sismica degli edifici", *Rischio sismico di edifici pubblici, parte 1 – Aspetti metodologici*, appendice n.1, CNR Gruppo Nazionale per la Difesa dai Terremoti.

Carper K.L., Feld J. (1997), *Contruction Failure (2nd ed.)*, Wiley J. & Son, Inc, New York.

Diamantidis D. (1987), "Reliability assessment of existing structures", *Engineering Structures*, 1987, Vol. 9, July.

Disegno di Legge n° 4339-bis, (1999), *Disposizioni in materia di regolazione del mercato edilizio e istituzione del fascicolo del fabbricato*, Senato della repubblica, XIII legislatura.

Deliberazione n° 166/1999 del Consiglio Comunale di Roma, *Istituzione presso gli edifici ubicati nel territorio del Comune di Roma di un fascicolo del fabbricato.*

Elms David G. (1999), "Achieving structural safety: theoretical consideration", *Structural Safety*, Special Issue, vol. 21, 313.

Fascicolo del Fabbricato per l'accertamento della consistenza statico-funzionale, Comune di Roma, 2000.

First Draft of Interim Technical Guidelines on the Inspection, Assessment and Repair of Building for the Building Safety Inspection Scheme (BSIS) (1997), Buildings Department, Hong Kong Government.

Interim Technical Guidelines on the Inspection, Assessment and Repair of Building for the Building Safety Inspection Scheme (BSIS) (1997), Buildings Department, Hong Kong Government.

ISO/FDIS 2394, Final Draft (1998), *General principles on reliability for structures*, ISO, Geneve.

ISO/CD 13822(rev) (1999), *Bases for design of structures – Assessment of existing structures.*

Micelli E. (2000), "L'Health and Safety file per la gestione in sicurezza degli immobili nel Regno Unito", *Convegno sul fascicolo del fabbricato*, Istituto Nazionale di Urbanistica, 22.03.2000, Roma.

Notes on the use of the Interim Technical Guidelines on the Inspection, Assessment and Repair of Building for the Building Safety Inspection Scheme (BSIS), (1998), Buildings Department, Hong Kong Government.

Pielert J., Baumert C., Green M (1996), *ASCE Standards on structural Condition Assessment and Rehabilitation of Buildings*, in Standards for Preservation and Rehabilitation, ASTM STP 1258, S.J. Kelly, Ed., American Society for Testing and Materials, pp. 126-136.

SCOSS (1999), *Structural Safety 1997-99: Review an recommendations*, in Twelfth report of SCOSS (Standing Committee on Structural Safety), London.

Stanghellini S. (2000), "Il fascicolo del fabbricato: non adempimento formale, ma vero strumento per la sicurezza e la manutenzione delle costruzioni", *Convegno sul fascicolo del fabbricato*, Istituto Nazionale di Urbanistica, 22.03.2000, Roma.

Structural Safety (Special Issue) (1997), *The Strategies and Value of Risk Based Structural Safety Analysis*, Vol. 21, n° 4, pp. 301-390.

Wicke Manfred (1994), "Assessment of bridges in Austria", *Evoluzione nella sperimentazione per le costruzioni*, CIAS, Merano.

Investigation and repair of an improperly constructed masonry barrier wall system

THOMAS M. KRAUTH
Vannoy & Associates, LLC, Silver Spring, U.S.A.

WAYNE T. RUTH
Masonry Solutions, Inc., Sparks, U.S.A.

DONALD W. VANNOY
Vannoy & Associates, LLC, Silver Spring, U.S.A.

BUILDING DESIGN AND BACKGROUND
The original design of a four story apartment building built circa 1995 in California, Maryland incorporated a composite brick and block design for the exterior walls. This design required the exterior walls of the building be constructed of an exterior wythe of brick masonry attached to a concrete masonry unit (CMU) back-up by means of galvanized truss type horizontal reinforcing and a solidly filled collar joint.

Wall section designs varied from location to location in regards to size of back-up block (4" or 6") although all sections called for an approximate two-inch solidly filled collar joint. This grouted collar joint was required structurally so that the exterior wall would behave as a composite wall. An additional function of the grouted collar joint was that of creating a barrier wall system in which the collar joint would act as a barrier to any water which penetrated the exterior wythe of brick masonry. Flashing and weephole details were shown on the original drawings at window and door heads and sills.

Subsequent to the construction of the building, water penetration problems began to occur primarily along the north and east elevations of the building. Wind-driven rains would routinely result in water penetration to the interior of the building. Water penetration was sporadic although often occurred at floor levels. In an attempt to remedy the water penetration problems, a siloxane penetrating water repellent was applied to the exterior face of the building. Although the water repellent appeared to decrease the amount of water penetration into the building, it did not completely eliminate it and further investigation into the water penetration problems was deemed necessary.

INVESTIGATIVE AND SAMPLING PROCEDURES AND FINDINGS
A preliminary visit to the property allowed discussions with building personnel, review of pertinent building plans, and a cursory inspection of the building. Additional investigative procedures and requirements were developed and an in-depth investigation into the water penetration problems was subsequently undertaken. Investigative techniques included destructive testing and exploration and material sampling. Several samples of mortar and brick masonry were acquired from various areas of the building, including a brick prism, mortar segments and individual brick units.

Forensic engineering: the investigation of failures. Thomas Telford, London, 2001.

Multiple wall deficiencies were observed during the investigative and sampling procedures. The deficiencies were all primarily the result of poor workmanship and resulted in a non-performing exterior wall system.   The following describes these deficiencies and their corresponding effects on the performance of the exterior wall assemblage.

Improperly Constructed Collar Joint
The brick prism removal location revealed a partially filled collar joint. Poorly consolidated collar joint mortar was present at the upper half of the prism area above a layer of horizontal reinforcing. This configuration evidenced placement of the collar joint mortar without regard to proper filling and consolidation, often the result of a practice known as slushing (filling joints by pushing mortar into the space with a trowel). Removed brick units revealed a completely omitted collar joint at least four feet in any direction.

The effects of the improperly constructed collar joint were two-fold. First, the structural integrity of the exterior walls was compromised. A solidly filled collar joint was designed to create a composite masonry structure. The solid collar joint was intended to bond the exterior brick and interior back-up block together and help generate composite action by transferring stress between the two wythes. In effect, the combined wythes act together as a single member. Without a solidly filled collar joint, this transfer of stress did not occur and the structural integrity of the wall was significantly compromised.

The second result of the improperly constructed collar joint was the compromise of the water resistance of the exterior walls. The solidly filled collar joint was architecturally designed as a barrier to stop the inward migration of wind-driven rain. The solid collar joint functions as a barrier to repel any water which passes through the exterior brick wythe. The omission of the collar joint resulted in unimpeded water migration into the unfilled collar joint and through the interior back-up block. Partially filled collar joints typically resulted in pockets where water collected and bridged to the interior of the building.

Incompletely Filled Mortar Joints
Removal of the brick prism and brick units at three separate locations revealed incompletely filled mortar joints. Incompletely filled mortar joints result in increased water permeability, reduce the strength of the masonry assemblage, and may contribute to disintegration and cracking due to water penetration and subsequent freezing and thawing.

Facial Separation Cracks
Widespread facial separation cracks were observed at each elevation of the building. Facial separation cracks are openings in the wall face between brick and mortar, usually 0.04 in. or less in width. They are most frequently caused by inadequate tooling of mortar joints during construction but may also be caused by thermal contraction of masonry units and mortar and less frequently by mortar shrinkage.

Brick Unit Cracking
Brick unit cracking was sporadically evident at the perimeter of the building. Brick unit cracks were typically straight and vertical and located at the center of the brick units. Brick unit cracks of the type evident at the building are typically caused by water penetration and subsequent freezing and thawing but may also be caused by inadequate quality control in the brick manufacturing process. Brick unit cracks result in an increase in the permeability of the masonry walls.

Clogged Weepholes
Mortar clogged weepholes were present at multiple flashing locations surrounding the building. Weephole installation was detailed in combination with flashing installations to properly drain any water collected on the flashing. Mortar clogged weepholes prevent the proper drainage of water from within the wall assembly and can result in water infiltration into the building and disintegration and cracking of the brick units.

MATERIAL TESTING AND ANALYSIS
Collected masonry and mortar samples were laboratory tested in order to detect any material defects and to verify specified mortar proportioning. The brick prism, five unsealed brick unit samples, and a sample of collar joint mortar were packaged and shipped to a testing laboratory. The laboratory was authorized to perform a petrographic examination of the mortar, a chemical analysis of the mortar, and determine the water absorption characteristics of the brick units.

Analysis of the laboratory test results revealed mortar proportioning and mixing was adequate. The petrographic and chemical analyses revealed mortar composition consistent with Type S mortar, as was specified for use on the project. The hardened mortar properties were acceptable although brick-mortar boundaries exhibited weak bonds. This was evidenced by easy fracture at brick-mortar interfaces and excessive depths of carbonation at bedding surfaces. Patchy accumulations of yellow-brown soils on the brick surfaces may have contributed to said weak brick-mortar bond and subsequent water infiltration. These accumulations evidenced poor handling of materials prior to use. Organic chemical analysis of the bed joint mortar did indicate the presence of minor quantities of glycerol or related diols. The presence of glycerol typically indicates the addition of anti-freeze to the mortar to improve workability in cold conditions or to act as an accelerator. The trace amounts of glycerol found in the bed joint mortar samples were minor and should have had no deleterious effects on the mortar. Water absorption testing of the five unsealed brick units indicated saturation coefficients and absorption characteristics acceptable for grade SW facing brick units, as were specified for use on the project.

PRELIMINARY REPAIR STRATEGIES AND TEST PANEL CONSTRUCTION
Based upon the above investigative and laboratory findings, the primary requirement for repairing the deficient wall system was the restoration of the structural integrity of the wall. In order to accomplish this goal, the empty collar joint must be solidly filled with a load bearing material as was originally intended. The second repair goal was the elimination of water infiltration into the building. This goal should also be met by the filling of the collar joint in that the barrier wall system originally intended will be restored.

The most effective method of accomplishing these goals is grout injection. This well researched method for repairing and strengthening masonry walls involves low-pressure injection of a fine cement-based grout into the collar joint. The injected grout is formulated to completely fill the empty collar joint and bond the exterior brick and interior back-up block together and restore composite action between the two wythes. In addition, the grout solidifies to a dense layer to prevent moisture passage to the wall interior while still permitting some level of vapor transmission through the wall system.

A test panel was constructed in order to test the grout injection procedure and performance. Prior to construction of the test panel, borescopic inspections of the building's internal wall conditions were conducted. The borescopic inspections further revealed extensive voids within the collar joint. Partially filled areas or "mortar patties" within the joint were observed in multiple locations although their locations did not appear to follow any pattern. These areas of partially filled collar joint complicate grout injection in that the grout may not be able to completely fill all voids and cracks within the mortar patties and that the patties can cause air pockets to form during the injection process. These expected complications can typically be addressed through grout injection procedures.

Based upon the desired material properties, performance criteria (i.e. strength, bond and waterproofing) and in-place conditions (i.e. cavity size and mortar patty size and location), a grout mix formulation was developed for use in the test panel construction. The test panel grout formulation, consisting of Portland cement, sand, fly ash, lime, water reducing admixture, and water, has the material properties shown in Table 1.

Table 1: Grout Material Properties

Gelman Water Retention (ml water collected)	< 2.0 ml
ASTM C940 Bleeding Test (5)	0
ASTM C939 flow (seconds)	8 to 14
ACE Flow (puddle diameter, inches)	7-1/4"
ASTM C1019 Grout Compressive Strength (psi)	3100 @ 28 days
Typical Shear Bond Strength Range (psi)	74 to 170

The formulated test grout was injected under pressure through ports into the collar joint cavity at an approximate 60 s.f. area at the south end of the building. Borescopic examination of this area revealed a primarily empty collar joint with small mortar patties located at the top of the test area. Prior to grout injection, expandable foam was injected into the collar joint on either side of the test area in order to provide an end dam to prevent unwanted lateral grout flow. A borescope was utilized to monitor grout flow through the cavity. The test grout injection successfully filled the empty collar joint, although as expected, additional injection ports and effort were required to inject the grout at the partially filled collar joint area.

FIELD ADAPTED ASTM E514 TESTING PROCEDURES AND RESULTS

In order to assess the performance of the proposed repair strategy with regards to water permeability, field adapted ASTM E514 testing was performed. The grout injection test panel was subjected to a field adapted moisture penetration test described in ASTM E514 before and after (four days to allow for grout curing) injecting the collar joint with grout. The test involved clamping a pressure chamber to the face of the wall and applying water at a rate of 40.8 gal/hr. The entire chamber was pressurized to 10 psf and the test was continued for a minimum of four hours. Information recorded during the test included water loss through the wall assemblage and appearance and location of moisture at the inner back-up or at brick units adjacent to the test area (lateral migration).

Results of the field adapted ASTM E514 tests showed that fully grouting the collar joint with the chosen grout formulation significantly reduced water penetration through the wall. Flow rates prior to grout injection were measured at 2.69 gal./hr. while post grout injection rates were measured at 0.24 gal./hr., equating to a 91% reduction in water penetration.

Empirical standards of performance have been developed through the accumulation of historical data for the Field Adapted ASTM E514 Test. These standards are presented in

Table 2. As can be seen, prior to grouting, the masonry wall assemblage would be rated poor in regards to service, materials and quality of construction. Subsequent to filling the collar joint by means of grout injection, the wall assemblage achieved a rating of excellent in each category.

Table 2: Field Adapted ASTM E514 Empirical Standards of Performance

Water Loss Rate (gal./hr.)	Service Performance Rating	Materials and Quality of Construction Rating
< 0.5	Excellent	Excellent
0.5 to 1.0	Expected	Standard
1.0 to 2.0	Questionable	Questionable
> 2.0	Poor	Poor

Source: Hoigard Et Al., "Including ASTM E514 Tests in Field Evaluations of Brick Masonry," Masonry: Design and Construction, Problems and Repair, ASTM STP1180

REPAIR IMPLEMENTATION
As a result of the effectiveness of the test panel construction with regards to restoration of the structural integrity of the wall and elimination of water infiltration into the building; grout injection was chosen as the appropriate repair method for the improperly constructed barrier wall system. A specialized masonry repair contractor was retained to effect the wall repairs through grout injection.

Pre-Injection Evaluation and Analysis
Due to the complexities involved in the grout injection process and the random nature of the masonry deficiencies at the building, additional evaluation of the entire structure (including masonry ties, flashing details, collar joint profiling, moisture infiltration locations, lintel conditions and expansion joints) was required prior to grout injection. The first stage of this evaluation required a personnel lift to reach the upper levels of the fully-occupied structure. Three-eighths inch diameter holes were drilled at mid-points in the head joints of the oversize brick into the collar joint area in order to allow for borescopic investigation of the wall system. These observation points were provided in the following locations around the structure:

- Parapets
- Heads and sills of windows
- Returns
- Changes in wall profile
- Sealant joints
- Apparent moisture infiltration
- Reported moisture infiltration
- Areas of efflorescence or staining
- Changes in designed wall width
- Window and door perimeters
- Walls adjacent to flashing
- Weep hole locations

Notably, facial separation cracks and improperly filled collar joints and masonry head joints were observed throughout. Facial separation cracks would be filled during the grout injection process by the injected materials moving to the face of the wall from the collar joint side. Improperly filled head joints would be filled from their backside when adjacent to the collar joint cementitious injected fill, or at locations of the drilled injection ports.

In addition to the borescopic observations, physical observations were made at the underside of the wood-framed roof, particularly at the intersection of the roof and the exterior wall. Observations were recorded as to the moisture infiltration frequencies and locations at each elevation, each floor, and each unit in order to compare and analyze the moisture related problems with physical conditions of the structure.

A pachometer was utilized to scan a representative portion of the building for joint reinforcement. The reinforcement was expected to be present at 16"o.c. vertically, as indicated in the construction documents. The presence of joint reinforcement could not be determined with the pachometer where structural steel was present. Such areas, which comprised approximately 20% of the region inspected, were inspected using a borescope. Of the regions where joint reinforcing could be inspected by pachometer, the total expected linear footage of reinforcing was 831 lineal feet, however the amount of joint reinforcing present was 394 lineal feet. This finding confirmed the need to augment the wall with ties prior to the cementitious injected fill process.

Compilations of the data from these investigations were analyzed to determine areas of potential anomalies and deviations from standard cementitious fill procedures. Of particular concern were areas of mortar entrapment and "patties" within the collar joint area. These areas needed to be identified and the protocol modified so that air entrapment during the injection process would be avoided.

Additional data was derived from the use of thermal transfer technology to delineate patterns indicative of collar joint voids. The thermal imaging was verified through pulse echo imaging and borescopic investigation. These methodologies indicated "mortar ledges" at apparent scaffold heights, repeating in different wall areas. This pattern is presumably related to either dumping of mortar at the end of the work day or perhaps a visit onto the scaffold from an inspector. These "ledges" needed particular care throughout the injection process so that air pockets would not ensue.

Acquired investigative data was provided to the cementitious injected fill technicians along with the injection protocol. The injection protocol established required mixes for injected fill that would be harmonious with the original materials. In some areas, different cementitious fill mixes were specified in order to accommodate void and fissure structures within the wall. The protocol also included a pattern of injection ports relating to the insitu wall conditions and established injection procedures including the removal and clearing of sealant joints, the preparation of flashing areas, and procedures to be used at changes in wall profile.

Installation of Cementitious Injected Fill
Complete injection of all facades of the building followed. A trained and certified crew of technicians with an average experience at grout injection of over five years, worked through winter conditions, from an enclosed and heated staging, while continually monitoring the interior of the fully occupied building. Random unannounced visits by the cementitious injected fill engineer insured quality control through flow tests, physical inspection, pulse echo imaging, and material property testing.

During grout injection, several challenges and anomalies were encountered. One such anomaly was discovered at a specific area of the building while analyzing the flow of the cementitious injected fill. A membrane flashing within the wall was discovered as being installed upside-down; that is, diverting water from its high-side in the brick down to embedment in the C.M.U. backup. Complete grouting of the cavity and encapsulation of the membrane flashing in this area served to address the deficient flashing by relying on the waterproofing integrity of a properly constructed barrier wall system.

Prior to injection, the technicians addressed those areas identified as requiring additional ties through installation of retrofit anchors. Sealant joints, cleared and removed for the process, were cleaned of all prior blockages and after injection, resealed. Special treatments were provided to eliminate leakage particularly at window head areas, and in areas with improperly installed flashing.   Furthermore, clogged weepholes were unclogged to ensure proper function.

After completion of the procedure, injection ports were re-pointed with compatible pigmented materials, and the walls and areas were cleaned of any minor spotting. Recommendations were provided to the owner of remaining roofing and window issues that needed to be addressed to effect a long term solution.

Finally, thermal imaging was again employed to verify the filling of all voids. Some small areas, a handful in all, were identified as a suspected void condition. These areas were re-injected with an appropriate cementitious injected fill, and re-tested to insure their integrity.

CONCLUSION
The original design of the subject building required the exterior walls of the building be constructed of an exterior wythe of brick masonry, a solidly filled collar joint, and a concrete masonry unit back-up. This system was required structurally to provide a composite wall which would transfer stresses between the two wythes and architecturally to create a barrier to stop the inward migration of wind-driven rain.   This is an acceptable wall system design and by all accounts was properly designed and specified.

Investigative procedures revealed multiple construction deficiencies which resulted in the non-performance of the masonry wall system at the building. Of primary concern was the omission of a solidly filled collar joint. This omission resulted in a compromise of the structural integrity and water resistance of the exterior walls.   Material testing and analysis revealed mortar and brick unit properties consistent with specified standards although further evidence of poor construction techniques was identified.

In order to properly repair the deficient wall system, the structural integrity of the walls needed to be restored. In order to accomplish this goal, the empty collar joint needed to be solidly filled with a load bearing material as was originally intended.   The most effective method of accomplishing this goal was grout injection. In addition to restoring the structural integrity of the walls, grout injection reduced the water permeability of the exterior wall system by restoring the barrier wall design and filling incompletely filled mortar joints, facial separation cracks, and brick unit cracks.

Field Adapted ASTM E514 testing in combination with test panel grout injection proved successful in filling the empty collar joint and reducing water penetration through the wall by 91%. The water permeability levels attained subsequent to grout injection rate excellent with regards to service performance considering minor infiltration is expected due to material absorption and occurred as a result of an air pocket within the collar joint.

The repair of the deficient exterior wall system was subsequently performed. Building-wide evaluations specific to the grout injection process were conducted using advanced techniques including fiberoptic observation, pachometer testing, thermal imaging, and pulse echo imaging. Acquired investigative data was provided to the injection technicians along with an injection protocol which established required mixes, injection port patterns, and injection procedures.

Although several challenges were encountered during the repair procedure, the improperly constructed barrier wall system was successfully repaired using grout injection. At the conclusion of the repair, thermal imaging verified that the collar joint had been completely filled, thus restoring the structural integrity and reducing the water permeability of the exterior wall system.

American	SI
1.0 s.f.	0.09290 s.m
1.0 in.	2.54 cm
1.0 ft.	0.3048 m
1.0 gal.	3.785 l

Assessing building wall system failures

KIMBALL J. BEASLEY, P.E. AND DOUGLAS R. STIEVE, AIA
Wiss, Janney, Elstner Associates, Inc.
1350 Broadway, Suite 206
New York, New York 10018-7799
USA

ABSTRACT
Failure of building wall systems can result from numerous causes. Uncontrolled water penetration, restrained differential movement, deterioration, or other influences can lead to unacceptable function of the building wall. Failure symptoms often include instability or collapse of walls or wall components or unacceptable serviceability, such as water leakage, cracking, or aesthetic problems. Underlying failure causes may be traced to deficiencies in design or construction, or a combination of the two. Clay masonry, natural stone, or other traditional wall materials may fail due to improper original workmanship or neglect of maintenance while in service. Contemporary wall systems with complex combinations of materials or configurations may fail due to material incompatibility or unforeseen loading or structural behavior. This paper will present a variety of wall system failures encountered by the authors and assessment methods used to investigate common failures. Well established investigation tools and techniques, as well as state-of-the art non-destructive testing, will be addressed. Conceptual repair methods for stabilization and repair of failed wall systems will also be included.

Over the past few decades, building walls have evolved from massive masonry building support elements to a much less substantial protective "skin" intended simply to keep the weather out and occupants in. Since these newer walls are usually subject to complex physical interactions with building components, and since there are narrower tolerances and greater performance expectations, the potential for failure is greatly increased.

Recent tools and techniques have been developed to investigate and resolve wall failure causes and to test and evaluate repair options.

Forensic engineering: the investigation of failures. Thomas Telford, London, 2001.

WALL SYSTEM CATEGORIES

Building walls may either be load bearing or non-load bearing. Traditional load-bearing walls often have a great deal of residual capacity to resist applied forces as well as the building's gravity loads. These load-bearing walls may be designed as solid barrier walls or cavity walls. They are sometimes made of brick or concrete masonry and clad with terra cotta, ashlar or dimension stone, precast concrete, bricks or cast stone.

Contemporary non-load bearing walls may be constructed with brick or concrete masonry veneer or with composite prefabricated wall panels of thin stone veneer, metal plate, Exterior Insulation and Finish Systems (EIFS), polymer-based sandwich panels, ceramic veneer on cementitious backer board or numerous other materials. These composite walls are usually much lighter than load-bearing walls with less redundancy against unanticipated forces. Further, architectural creativity or simply economic pressure often inspires novel, unproved combinations of materials or wall configurations.

WALL FAILURE CATEGORIES

Walls fail when they collapse or shed pieces (stability failures) or when they leak, deteriorate, crack, or stain (serviceability failures). Some failures may be tolerable and others may be catastrophic. According to the Merriam-Webster Dictionary "failure" means "...the inability to perform a normal function adequately." There is no question that a severely cracked or displaced wall has failed, however, a wall with a latent failure -- one that looks intact but cannot sustain normal loads because of missing or damaged hidden supports -- can be a far more troubling failure because it is difficult to detect and can collapse without warning.

Water Penetration

Water that penetrates the wall surface usually migrates downward within the wall. Leakage to the interior is avoided by solid "barriers" that inhibit the water flow or by "cavities" that convey the water down to internal flashings that should be watertight with weepholes to drain water collected on the flashing. Leakage is likely to occur if the barrier walls are not completely solid or if the wall cavities are blocked, or if flashings are breached or poorly positioned.

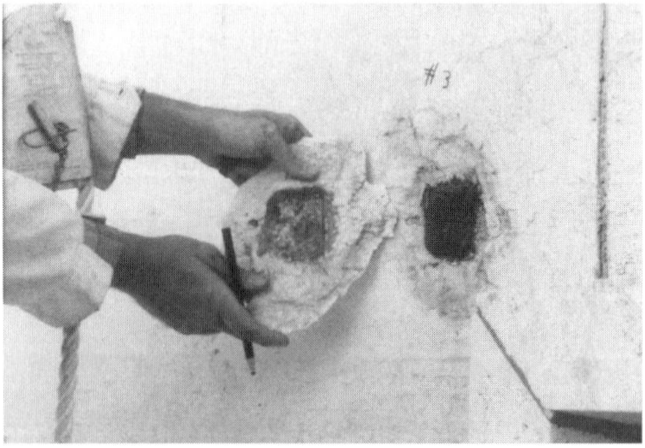

Figure No. 1 – Thin stone veneer has spalled from corrosion jacking of embedded reinforcing steel.

Corrosion Jacking

When embedded mild steel reinforcement or support members corrode, the oxidation product (rust scale) expands up to 10 times the original steel volume. Steel corrosion initiates from sustained water exposure and from deleterious chemicals such as chlorides. Figure No. 1 shows stone veneer spalling caused by corrosion jacking.

Freeze-thaw Deterioration

Stone, concrete, mortar, or clay masonry wall materials can deteriorate from freeze-thaw action when saturated with water and subjected to cyclic freezing and thawing. This phenomenon occurs when water fills microscopic voids in the material and expands upon freezing, progressively rupturing the adjacent masonry. Figure No. 2 shows a concrete wall element severely damaged from freeze-thaw degradation.

Figure No. 2 – Architectural concrete severely deteriorated from freeze/thaw action.

Thin Marble Hysteresis

In contemporary wall construction marble cladding panels are often cut into thin (2 to 5 cm) slabs. Certain types of marble have been known to experience failure in the form of bowing, disaggregation (sugaring), and strength loss when cut into thin panels and exposed to the weather. Testing and laboratory studies have shown that this behavior is related to marble's anisotropic properties and complex grain morphology (interlocked hexagonal shaped calcite crystals bound together with a thin calcite binder). Thermal cycling leads to intergranular fracturing along the grain boundaries; the grains tend to dislocate and not return to their original position [Ref. No. 1]. This process is termed hysteresis. This phenomenon has resulted in several recent spectacular and costly failures in the United States and abroad.

Restrained Differential Movement

Restrained accumulated expansion and contraction of the wall or facade material relative to the building's structural frame is a frequent cause of wall failures. This condition can be caused by thermal expansion of the facade or wall, coupled with resistance from the structural frame. The structural frame moves relatively little because it is usually sheltered from the weather and consequently wide variations in temperature. Also, slow moisture expansion of fired clay masonry products or moisture shrinkage of concrete masonry products contributes to potential wall failure from differential movement between the facade and the underlying structure [Ref No. 2]. Load-induced elastic deformation and shrinkage and creep of reinforced concrete frame construction further increases vertical compressive forces in the cladding by shortening the structural frame. Differential movements accumulate in long or tall walls, resulting in stress levels in the wall many times greater than from normal gravity or wind loading.

Figure No. 3 – Prefabricated thin tile-faced composite wall panel failed from moisture movement of the underlying wood fiber-reinforced cement board.

Figure No. 3 shows a ceramic tile veneered, prefabricated composite wall panel that failed from moisture-loss shrinkage of the wood-fiber reinforced cement backer board. The resulting differential tile-to-backer board movement resulted in eccentric planar forces that caused the thin (about 2 cm) composite wall panel to bow and separate from the wall's metal stud support frame.

WALL FAILURE INVESTIGATION SCOPE

Certain systematic approaches have proven successful in helping the investigator to recognize symptoms and patterns of distress to identify the source or character of wall failures. A proper understanding of the failure mechanism will then help in devising effective methods of correcting the resulting damage as well as resisting further failures.

Wall failures as well as building failures in general are usually investigated: (1) to determine appropriate repair procedures, (2) to further the body of engineering knowledge to avoid similar failures in the future, and (3) to help assign responsibility for the economic loss.

The scope of the failure investigation usually depends on the nature, severity, and consequences of the failure [Ref. No. 3]. Wall failure investigations should involve a systematic, logical approach of collecting and analyzing data. The investigation should not be reduced to a "cookbook" or "check list" procedure that discourages the free thought, imagination, and flexibility essential to any failure investigation [Ref. No. 4].

The wall failure investigation may involve: (1) acquisition of data, (2) analysis of data, (3) development and evaluation of hypotheses, (4) formulation and communication of opinions.

WALL FAILURE INVESTIGATION METHODOLOGY

Initial Site Visit

The initial site visit objective will depend on the type of wall failure. Rapid data collection will be necessary where a collapse and associated rescue and cleanup effort is underway. Photography (still and video cameras) can be used to quickly capture the position of remaining building elements or collapse debris. A digital camera with a laptop computer can collect and transmit great quantities of graphic information quickly. However, in cases where litigation is anticipated, it is possible that digital photographs may not be allowed as evidence by the courts because the images can be manipulated. If field conditions will not be disturbed for some period of time, the initial site visit may be delayed until after the document review phase.

Document Review

Relevant information may be acquired from: (1) original structural and architectural drawings, (2) specifications, (3) submittals including shop drawings or other correspondence produced during original construction, (4) prior maintenance or repair records, (5) interviews with witnesses, and (6) observations in the field. Detailed field inspection notes, in-situ testing results, and laboratory test data are common sources of investigation information.

Visual Condition Survey

Field observation of wall distress using binoculars and telephoto equipment can be documented by annotations onto building elevation drawings or onto high-resolution photographs. The superimposed conditions observed can be viewed over an entire elevation to help identify the patterns and character of deterioration and distress. These surveys can also establish a baseline for subsequent detailed inspections and determine the potential scope and location for future repair work.

Detailed Inspections

As-built construction and the conditions likely to have existed prior to the failure need to be determined via detailed close-up inspections. Such inspections will usually involve measurement of building element dimensions and positions as well as examination of subsurface wall elements, facade connections, and adjacent construction via probe openings or other inspection devises (see Wall Failure Investigation Tools, below).

Sample Collection and Custody

Laboratory examination and testing of material samples is sometimes part of the wall failure investigation. Samples of deteriorated material and companion samples of intact elements for control testing may help to isolate the cause of deterioration. The quantity, locations, and type of samples taken will be based on the nature of the failure. To be statistically representative sample locations should usually be reasonably random.

Organization and Communication of Finding

The wall failure investigation report may take a variety of forms. Voluminous, highly technical reports usually require an executive summary. Reports can be brief letters with appendices of subspecialty studies (e.g., metallurgical, petrographic). Attorney clients may request an oral report to assess the culpability of their client before the information is commemorated in writing. The information in the report should follow a logical sequence from factual findings, to opinions, to recommendations: most substantial wall failure investigation reports contain sections on introduction and background, observations and factual information, analysis and discussion, opinion/conclusions, and recommendations. Relevant assumptions and limitations due to incomplete data which formed the basis of the report should be described. Depending on the client's needs, communication of investigation findings may also include presentations or development of models and court exhibits.

WALL FAILURE INVESTIGATION TOOLS, NON-DESTRUCTIVE TESTING

Deteriorating building elements do not always exhibit obvious symptoms. Corrosion behind the wall surface may advance to the point of failure with little warning. Non-destructive testing methods provide a means of evaluating the material or hidden elements with little or no damage.

Figure No. 4 – A scratch gage is used to detect movement across a building joint.

Measuring Devices
Simple measuring devices include a tape measure, plumb bob, or optical crack comparitor. More sophisticated measuring devices, such as a theodolite is used to measure precise relative positions of visible building facade areas relative to a reference point. Cracks and joint movement gages involve calibrated plastic scales, scratch gages (Figure No. 4) or sensitive dial gage devices. High precision global positioning systems (GPS) and various electrical transducers and accelerometers can be used to accurately monitor building movements.

Borescope
The borescope employs fiberoptics to view hidden spaces. The borescope's small diameter (6 or 8 mm) metal tube is inserted through a small hole or joint (Figure No. 5). The device includes a power supply and light source, as well as an eye piece and camera mount. Both still or video camera may be used with the borescope.

Metal Detector/Pachometer
The location and size of embedded steel reinforcement and steel support or anchor elements within the wall can be determined with sensitive metal detectors of pachometers. The pachometer measures changes in electric inductance to indicate the presence of underlying ferromagnetic metal. The intensity of the electromagnetic signal is used to precisely locate the steel.

Rebound Hammer
The rebound hammer, also called Swiss or Schmidt Hammer, roughly correlates compressive strength of concrete or masonry based upon the rebound of a mass on a spring after impacting the wall surface. The rebound displacement is indicated on a calibrated scale. The rebound hammer provides a quick determination of the relative compressive strength test from a concrete core or masonry prism. The test results can be used to extrapolate approximate compressive strength for other areas of the structure. The rebound hammer is generally used to supplement but not replace compressive strength testing of concrete cores or masonry prism.

Pulse Velocity Methods
The velocity and waveform of a sonic pulse traveling through concrete or masonry approximates the masonry's relative strengths or to identify internal discontinuities. The technique was originally developed for concrete and is still considered experimental for use with masonry. A pulse of ultrasonic energy is transmitted into the concrete or masonry and a receiving transducer detects the transmitted signal. The pulse velocity is determined by

dividing the distance between the transducers by the pulse transit time. The velocity of a sonic pulse corresponds to the material's elastic modulus and strength.

Strain Relief
One method of strain relief testing involves application of carbon-filament strain gages to the surface of the subject wall element. The instrumented unit or area is then isolated from the surrounding wall by saw cutting. The strain value before and after saw cutting is measured. Using the modulus of elasticity of the base material, the in-situ residual stress is then computed. This technique is particularly useful to determine the potential for future compression or buckling failure of masonry facades and for cracking of wall panels.

LABORATORY ANALYSES

Petrographic Microscopy
Petrographic analysis involves a standardized microscopic examination of stone, concrete, brick, or mortar based on the method outlined in ASTM C856, *Petrographic Examination of Hardened Concrete*. The objectives of a petrographic examination are often to determine such conditions as: (1) the presence of microscopic defects, (2) visible indicators of deterioration, (3) evidence of unsound or reactive aggregate, unhydrated cement, carbonation of the concrete and mortar; and (4) the cement content, water-cement ratio, percent of entrained or entrapped air, characteristics of the air void system, and the degree of consolidation of concrete or mortar. In general, petrography indicates the overall quality and soundness of the stone, brick, concrete, or mortar.

Chloride Content
The chloride ion content in concrete or mortar provides quantitative evidence of the potential for corrosion of embedded steel elements. The chloride content is determined by an acid-digestion, potentiometric titration procedure.

Mechanical Tests
Properties of materials, such as compressive or tensile strength, modulus of elasticity, etc., are obtained to determine physical characteristics of wall elements. With masonry walls, coupons or prism samples can be saw cut from the wall and tested to failure in the laboratory.

Freeze-Thaw Testing
Masonry and concrete materials susceptibility to cyclic freezing and thawing is measured by exposing uniform sized samples to temperatures above and below freezing while critically saturated. Deterioration is determined by observing the samples for cracking or other forms of deterioration and weighing the samples periodically during the test to assess weight loss related to fragmentation. Dynamic modulus methods determine variations in the sample's resonant frequency as an early detection of internal sample degradation.

REPAIR CONCEPTS
Depending on the nature of the wall system failure, temporary repairs or permanent repairs (or both) will need to be designed.

Temporary repairs

Temporary stabilization of deteriorated walls may be implemented on an emergency basis. Such repairs may involve netting, strapping, shoring, or anchoring unstable facade elements (Figure No. 6). Removal of loose or dislodged pieces will often be undertaken as emergency protection measures. Consideration for aesthetic impact is superceded by the need to avoid falling pieces. Sometimes sidewalk bridging is erected to protect the public; however, several months or even years may pass before permanent repairs are implemented. Damaged facade areas need to be protected from the elements if permanent repairs are delayed.

Permanent repairs

Long-term repairs may include replacing damaged materials, securing cladding elements, or complete wall recladding or overcladding. Deteriorated or dislodged wall components may be replaced with similar materials or substitute materials. The decision to use replacement materials rather than to replicate in kind may be driven by economic or scheduling concerns. Preservation requirements, structural limitations, and fire codes may also restrict the use of replacement materials.

CONCLUSIONS

Wall system failures often occur because of poor construction practices or because the design is deficient or invites construction errors. The lack of built-in redundancy, either with the wall support or attachment mechanism or with the internal water drainage system, means that failure of a single wall component can lead to serious loss of function or even failure of the entire wall system. Both simple and sophisticated investigation tools and methods have been developed to effectively detect and evaluate wall system failures.

REFERENCES

Ref. No. 1 - Wildhalm, C., Tschegg, E., and Eppensteiner, W., "Acoustic Emissions and Anisotropic Expansion when Heating Marble," *Journal of Performance of Constructed Facilities*, American Society of Civil Engineers, New York, February 1997.

Ref. No. 2 - Beasley, Kimball J., "Masonry Cladding Stress Failures in Older Buildings," *Journal of Performance of Constructed Facilities*, American Society of Civil Engineers, New York, November 1988.

Ref. No. 3 - Beasley, Kimball J., "Failure Investigation, " *The Construction Specifier*, April 1998.

Ref No. 4 – Beasley, Kimball J. and Patterson, David S., "The Building Envelope," *Forensic Structural Engineering Handbook*, McGraw Hill, New York, 2000.